I0110682

A Sapiens Conundrum

Surviving Our Survival Instincts

David Walter

Wisdom
Editions

Minneapolis

**Wisdom
Editions**

Minneapolis

Second Edition April 2022
A Sapiens Conundrum. Copyright © 2021 by David Walter.
All rights reserved.

Published by Wisdom Editions, 6800 France Av S, Suite 370, Edina, MN
55435. No parts of this book may be used or reproduced by any means,
graphic, electronic, or mechanical, including photocopying, recording, taping
or by any information storage retrieval system, without the written per-
mission of the publisher except in the case of brief quotations embodied in
critical articles and reviews.

10 9 8 7 6 5 4 3 2

ISBN: 978-1-960250-74-2

Cover and book design by Gary Lindberg

Table of Contents

A Sapiens Conundrum

Surviving Our Survival Instincts

God blessed them and said to them, "Be fruitful and increase in number; fill the earth and subdue it. Rule over the fish in the sea and the birds in the sky and over every living creature that moves on the ground."

Genesis 1:28

… humanity's increasing destruction of nature is having catastrophic impacts not only on wildlife populations but also on human health and all aspects of our lives.

Living Planet Report 2020, World Wildlife Fund

The most decisive mark of the prosperity of any country is the increase of the number of its inhabitants.

Adam Smith, *The Wealth of Nations*

… the power of population is indefinitely greater than the power in the earth to produce subsistence for man.

An Essay on the Principle of Population, Thomas Robert Malthus

Nations with higher GDP per person have more of almost everything – bigger homes, more cars, higher literacy, better health care, longer life expectancy, and more Internet connections. Perhaps the most important question in macroeconomics is what determines the level and the growth of GDP.

N. Greggory Mankiw, *Macroeconomics*

CO_2 emissions per head have been strongly correlated with GDP per head. As a result, since 1850, North America and Europe have produced around 70% of all the CO_2 emissions due to energy production, while developing countries have accounted for less than one quarter.

Stern Review, "The Economics of Climate Change"

Chapter 1

A Biological Conundrum

Galileo Galilei caught a lot of flak for being scientific. He made observations of the sky through his telescopes and those observations gave him a different perspective on the earth's place in the solar system. He knew this would come with a cost, for his contemporaries were persecuted for professing the same beliefs. Common understanding of the earth's place in the heavens and the view of the Catholic Church at that time was geocentric: the sun and all other planets revolved around the earth. Based on his observations, Galileo, and others before him including Nicolaus Copernicus, believed in a heliocentric heavenly system: the earth and other planets rotated around the sun.

In 1633 the Church demanded that Galileo disavow his beliefs. It may be a myth, but the story goes that when arising from his knees after kneeling to recant in front of a jury of Catholic cardinals, Galileo said, under his breath, "*E pur si muove* (and yet it moves)."

This is not the only instance when scientific observations of the natural world contradict commonly accepted beliefs, strongly held. Before the voyages of exploration in the late 1400s and the 1500s, many assumed the world to be small, and flat. When Charles Darwin proposed in his book *On the Origin of Species* in 1859 the concept of gradual, incremental evolution of organisms over long periods of time, it was generally held that the earth was 8000 years old. In the

1840s, Dr. Ignaz Semmelweis observed an extraordinarily high rate of puerperal fever in a Vienna hospital maternity ward, and proposed that doctors sterilize their hands before assisting with childbirth, an idea that other doctors resisted. Several geologists, including Alfred Wegner in the early 1900s and Alex du Toit in the 1930s, proposed the concept of plate tectonics, that the earth's continents move, an understanding that ran counter to the opinion that continents were fixed in place.

In each of these examples, people resisted accepting the new concept. In some cases, severely. Now, it's common knowledge that our earth rotates around the sun, that life on earth evolved over millions of years, that the earth is quite large and somewhat spherical, that many diseases spread by the transfer of microorganisms and that continents drift. We face a Galileo moment now. We – anyone who is concerned with the state of the environment and all who are upset with the many facets of disharmony within our culture – are facing a contradiction between science-based observations of the natural world and commonly held convictions. Two convictions in particular are relevant: that technological innovation will help us solve problems we face and that population growth and economic growth are healthy for us. It does not really matter where these beliefs started; it is important that we are basing our decisions and public policy on them, and counting on them to improve our future. Are we justified in these policies? Are they helping us?

Questioning assumptions we're basing our public policies on is relevant because many problems humans face now are not getting better, they're intensifying. Instead of being resolved, many societal problems have become incorporated into our daily lives and effectively accepted until they erupt into a crisis. It makes sense to look deeply and explore what's really causing the systemic issues we face. If that means questioning policy assumptions that undergird our policy decisions, we are better off for those questions. So far, our policy decisions haven't been effective at resolving major issues.

Homo sapiens is a species in a biological conundrum. We are struggling in two realms – in the impacts we have on the environment and in the functioning of our culture – in ways that should not be possible, and in ways no other species struggles. Many biologists believe the problems we see are caused by a population that has grown too large. But in the natural world, there are inherent systems that should keep a population from growing large enough to cause these problems. Humans are the only species on earth with these environmental and cultural problems.

Environmentally, the two areas of concern are pollution and resource depletion. Our use of fossil fuels for the generation of electricity and for transportation generates carbon dioxide, methane and nitrous oxides in quantities that change the dynamics of the atmosphere and trap solar radiation, causing global warming (IPCC 2014, 2-8; World Meteorological Organization 2018, 1-8). With global warming, the oceans are becoming more acidic, are increasing in temperature and rising (IPCC 2014, 2-8; Cazenave 2010, 145-168, Hansen 2006, 14288), coral reefs are undergoing severe bleaching (Hoegh-Guldberg et al 2007, 1737-1742; Caldeira 2003, 365), glaciers are melting world-wide and the poles are losing ice mass (Dyurgerov and Meier 2005; ACIA 2005, 3; Hansen 2016, 532). Agricultural irrigation is causing aquifer and river depletion (Postel 1999, 3-12; Postel 2000, 941-942). The quantity and quality of water sources around the globe are in some areas proving to be marginal to supply regional populations (Nash 1993, 25-39; Cooley 2014, 1-6). Surface water is being polluted with chemicals that may be causing endocrine disruption in our bodies. (World Health Organization 2012, vii-xv). Our land development is generating species extinction at extremely high rates (De Vos 2014, 453-460; Leakey 1995, 232-235, 241, 244-245). The creation and use of plastics causes vast gyres in oceans and drifts of plastic washed up on beaches (UNEP 2005, 3-6; Ryan et al). Fertilizer applied to agricultural fields runs off and flows to oceans, causing large estuary dead zones (Diaz 2008). There are chemicals in our air, land and water – biological contaminants, nitrates, persistent organic pollutants, metals, industrial pollutants, radionuclides – that are toxic and make their way to our food and water and to our bodies (van der Leeden 1990, 496-503; McGinn 2000, 5-10; Nash 1993, 25-32).

Culturally, regions and countries are struggling with elements of dysfunction. Overall, we operate well enough to get by, but a tremendous amount of personal and governmental capital is spent on struggling with and trying to fix cultural dynamics that cause distress to populations. Inequality has been recognized as a problem for a long time (Wirth 1938, 9-18: Hayden 1995, 20-23). It is a simple concept but has outsized consequences in how it manifests itself in cultures. Inequality takes many forms, including inequality in access to education and to health care (Neckerman 2007, 335-347; Wilkinson 2010; Deaton 2013, 1-46), inequality in access to financial resources (Alvaredo 2013, 3; Alvaredo 2017, 8-20), and inequality in income and wealth (Piketty 2014). Racism and sexism negatively affect people's lives in some areas of the world. We have congested cities that may generate depression and anxiety in people (Ferrari 2010; Hidaka 2012). Some countries have high levels of incarceration (Crow 2011, 36-37). Governments, necessary to manage the functioning of countries, are often sources of economic, humanitarian and political dysfunction (Brennan Center for Justice Conference Report 2014). Some authors believe these cultural problems or ones parallel to them have been with human populations for hundreds of years, perhaps for thousands of years (Armelagos 1991; Wilkinson 2010, 3-4; Alacevich 2018, 13-14).

The trends of environmental degradation and cultural dysfunction are not getting better, they are getting worse, in spite of significant intellect, political will and personal energy applied to remediate them. This is a sharp wake-up call for us. We are the only species in the world to have a long-term, wide-spread, detrimental impact on the environment, and a culture that struggles to generate harmony to the degree we do. Business as usual is not working for the long-term health of our species. These trends are not happening by chance, and they are not at all consistent with the naturally occurring patterns of the earth's ecosystems. In our own self-interest, it behooves us to pay attention and be willing to look deeply at what is happening. *Homo sapiens* is a biological organism; every human needs clean air, food and water to be able to survive. We are a social organism; we all do better when we cooperate and maintain harmonious relationships with others in our population. The degradation of the resources we need and the dysfunction of the social fabric we live in do not bode well for the long-term viability of our species. Denial that these are pressing problems is common but that denial is itself a problem and does not help in any way.

Unaddressed, these problems could cross a tipping point and lead to a societal collapse, a breakdown of our human culture. Societal collapse has happened many times in the history of humans. Complex populations in Mesopotamia, south Asia, south east Asia, China, Greece, Central America, South America, North America, Africa and Greenland have all suffered collapses (Middleton 2017, xv; Railey 2008, 1-2). Specific causes for each collapse are not necessarily obvious, and are probably a function of multiple factors, but societal collapses do not happen by chance, and there is no reason to believe a collapse could not happen again. Collapses in the past were of regional populations; now, a risk is that with our large global population and the interconnectedness of world cultures and economies a collapse could be wider and deeper, involve more people, and consequently be more devastating.

Biologically, these problems should not be happening. It is a conundrum. In the natural world, the populations of every species of organism are constrained in size; there are inherent biological systems in place to keep the number of individuals in check, limiting the population size and precluding severe environmental or cultural damage. And yet *Homo sapiens* has gotten past those constraints. We are having deep and long-lasting impacts on the earth and are generating a culture that in many ways is unhealthy for us. What is going on?

These observations and this question are the starting point for examining how and why humans are capable of growing our population beyond what would naturally be possible and having a unique impact on the environment and our culture. This examination is important. If we can understand how humans have come to the spot we are in now we can use that knowledge to reverse the negative trends and adjust our position to be healthier.

Problem-solving

Humans are skilled problem-solvers. On both individual and species levels, when an issue emerges in front of us we tend to apply our intellect and skills to figure out a path to resolve the problem. Our history is filled with technologies we have developed and policies we have instituted to solve problems that were having an impact on us. We have created agricultural systems to feed us, a vast medical industry to keep us healthy, transportation infrastructure systems to allow people and products to move locally and around the globe, multiple energy systems, housing structures for us to live and work in and communication systems to connect us.

Historically, in many cases we have often taken a band-aid approach and worked to figure out what we can do to relieve the immediate pressure from a problem without questioning or trying to remedy the fundamental cause. For example, in the 1960s rivers in the U.S. caught fire: flammable pollutants had been discharged into waterways to a degree that they caught fire and burned. Legislation passed in 1972 that amended the Water Pollution Control Act of 1948 served to regulate dumping of pollutants into U.S. waters. The legislation had a positive impact on the level of pollution in the nation's waters – rivers no long burn – but it did not confront the larger question of what dynamics allowed that pollution to be present in the first place. Dumping waste into rivers had been occurring for a long time but never to the degree that was experienced in the 1960s. What was going on at that time that generated such pollution? That rivers were polluted to an extreme degree was not by chance. Burning rivers rose to the level of awareness that demanded a response, but not to the level of questioning the dynamics of how we got to that point. Rivers and lakes in the U.S. are still used to dump industrial waste, but to a lesser degree than previously.

Seeking to understand a large-scale, complex dynamic such as what allows dumping our wastes into rivers calls for more than managing the specific problems that emerge. The larger concern is that the problems exist at all. Before a certain point in time, those problems did not exist; after that point, they do; why? If we do not address the factors that cause the problems, they're still with us, even if ameliorated to a point that their impact is not as obvious or not as severe. If we want to confront our problems, really confront them, we need to get to core, foundational reasons for the existence of these issues that impact us. Long-term, macro-scale problems do not appear by chance – something is causing them.

A tremendous amount of effort and good intention have been applied to remediate the environmental and cultural problems we have now, yet the negative trends still continue. Are our problem-solving abilities being put to good use? Are we seeing the problems of environmental degradation and cultural turmoil from a perspective that is effective? Could there be a better way to view these problems? Are we asking the right questions? We, as a species, as regional populations, as communities, as individuals concerned about our health and the health of future lives, have a vested interest in a healthy environment and a healthy culture.

Drivers

Problems can be understood best by asking questions that frame problems clearly and that aim to penetrate deeply into them. With the framework of helpful questions, it is possible to tease apart the factors that create the problem, and in doing so find answers to how and why the problem exists. With such an understanding, paths to resolving these problems should become more evident.

This book seeks to discern the underlying, foundational drivers of the problems the human population is living with today. The questions asked here are: What is driving *Homo sapiens* to have long-term, negative impacts on the environment? What is driving the generation of dysfunction of our culture? Simple questions, but they aim for the core dynamics of what is causing the environmental and cultural problems we are experiencing.

Chapter 2

The Biology of Population Regulation

Humans are biological organisms. Our survival is dependent on basic biological needs: we would perish without clean water, food and air. To improve our odds of survival nature has endowed us with biological and psychological motivations: when hungry we seek food, when thirsty we seek water, our breathing functions autonomously. Every organism needs to absorb resources to maintain cell metabolism and therefore sustain life and every organism seeks to avoid disease and being consumed by another organism. The natural, biological world at its core is a simple and ruthless field of survival.

In this book, the word "natural" means an ecosystem or biological system unaffected by humans. The point is not to imply that every human impact is deleterious but to recognize that humans have had an outsized impact on the natural world through their development processes and those processes have affected the functioning of some ecosystems. "Natural" refers to biological systems without human influence.

Limiting Factors

Biologically, the vast majority of life forms on earth are based on daily solar input. Organisms called "primary producers," most of whom photosynthesize, are the foundation of life; these organisms synthesize the chemicals their cells need to metabolize by using light energy from the sun or in a few species chemical energy. On land, the vast majority of primary producers are plants; in water they are mostly algae. Plants and algae therefore form the base of the food chain of life on earth. The next level in the biological system are herbivores, organisms that consume plants. Herbivores cannot photosynthesize so cannot gain energy directly from solar radiation but they can gain energy by consuming plants. After herbivores are carnivores, animals that consume herbivores. Carnivores do not consume plants directly, but they are reliant on plants and the daily solar input that supports herbivores. This simplified explanation of the structure of biological life forms shows that the energy that all organisms need starts with solar radiation. We are all dependent on the sun for life.

In the natural world, an amazing system is in place that functions both to support and to constrain the population size of every species on earth. Population dynamics are complex, but simplified, the environmental variables supporting and constraining every organism are called 'limiting factors' of which there are three broad categories: resource supply, disease and predation.

Resource Supply

Every living organism needs to harvest resources to provide its cells the elements needed to maintain life. Different species require different resources. Temperature, water, altitude, sunlight, different gases in the atmosphere, soils, nutrients and food are all elements of the environmental resource base that every organism gets to work with – if the resources an organism needs are available, it can live. If the resources an organism needs are not available, it relocates or perishes. Fish need different resources than plants which need different resources than birds which need different resources than microbes.

One characteristic of all resources is that they are scarce: no resource on earth is in unlimited supply and supply of any given resource varies from region to region and changes over time. The size of an organism's population is in part dependent on the availability of the resources an organism needs to live: if abundant, the population is well-supplied; if scarce, the organism has a smaller population. In this "supply of resource – demand by organism" relationship, the population size is therefore limited by resource availability. One nuance: we can speak of "resource supply" as a way to indicate the supportive quality of resource availability or we can speak of 'resource scarcity' to indicate that resources are limited and constraining. Resources are effectively scarce and available in varying degrees, simultaneously.

Besides food resources needed to maintain life other resources are necessary for organisms to live. Many organisms have evolved to need particular environmental factors in their habitat: some woodpeckers need a specific species of tree for nests; some organisms require particular landscape to hide from predators; some predators have specialized in preying on one species of prey; some organisms struggle when their ambient temperature gets too cold or too warm. Many aspects of the environment qualify as resources organisms need to be able to live. All aspects of the environment an organism lives in impacts it to some degree; its response to the net effect of all these aspects is a function of its unique natural history.

Disease

The second limiting factor category is disease which often comes in the form of a microbe. The natural world contains a mind-boggling number of bacteria, viruses, protozoa, rickettsiae, fungi and worms that qualify as disease-causing organisms to humans (Barnes 2005, 2-6) and they each have a unique mechanism of interacting with their host and effecting a malady. A disease microbe is one whose life cycle involves gaining resources by infecting a host, who becomes ill or dies when invaded. Most organisms are vulnerable to some type or types of disease.

Human's relationship with disease has changed as our culture changed. Our contact with disease used to be mostly a function of geography – like other organisms, disease organisms and humans tended

11

to live in specific conditions, based on location, and we would be affected by those diseases we interacted with. That dynamic changed as we domesticated animals and developed trade – animals brought us into contact with some microbes that proved to be diseases to us; and trade brought long-distance transfer of disease organisms from one region to a distant one. With dense settlements and cities, some diseases found niches within our populations they would not have without that density. As we learned about causes of disease we developed medical science and medical technology that worked to minimize the impact of diseases on people. Currently, our ability to travel long distances allows disease organisms to move to new habitats quickly (Barnes 2005, xi-8).

Predation

The third type of limiting factor is predation. In the drive to harvest resources some organisms have evolved to directly consume other organisms as food. Blue whales prey on krill, deer prey on plants, bears prey on small mammals and plants. Prey species are therefore subject to predation pressure: the loss of population size due to being consumed by another species.

Like many other types of biological categories, defining 'predation' can be difficult. Keeton provides a broad definition: a predator is "… an organism that is free-living and that feeds on other living organisms" (Keeton 1972, 665). This definition helps understand predators relative to parasites, which tend to live in close proximity to hosts. Predators and prey have dynamic relationships; some predators can have the effect of decimating prey populations to the point of extinction, while some predators are more of a nuisance to prey species, such as the relationship of biting flies preying on mammals. Some predator/prey relationships benefit prey populations by culling weak or ill individuals. Predator/prey populations tend to be stable in long-term relationships, coming to an equilibrium that allows each to survive (Keeton 1072, 665).

In all natural ecosystems diverse types of organisms interact in overt and subtle ways and have the effect of supporting and constraining each other. Any single organism is naturally embedded in a biological system including its resource supply, disease organisms and predators; that organism is impacted by all three limiting factors simultaneously. The degree of abundance of supply of needed resources defines whether or not the organism has what it needs to survive. The limiting factors of disease and predation define the degree to which the population is weakened or decreased. The health and balance of the earth's ecosystems are maintained in part by the populations of each organism not growing too large – a population larger than what is normal in its ecosystem would have an outsized impact on other organisms. In any natural ecosystem at any given moment the population size of a species is balanced with all elements of the ecosystem, in large part through limiting factor pressure.

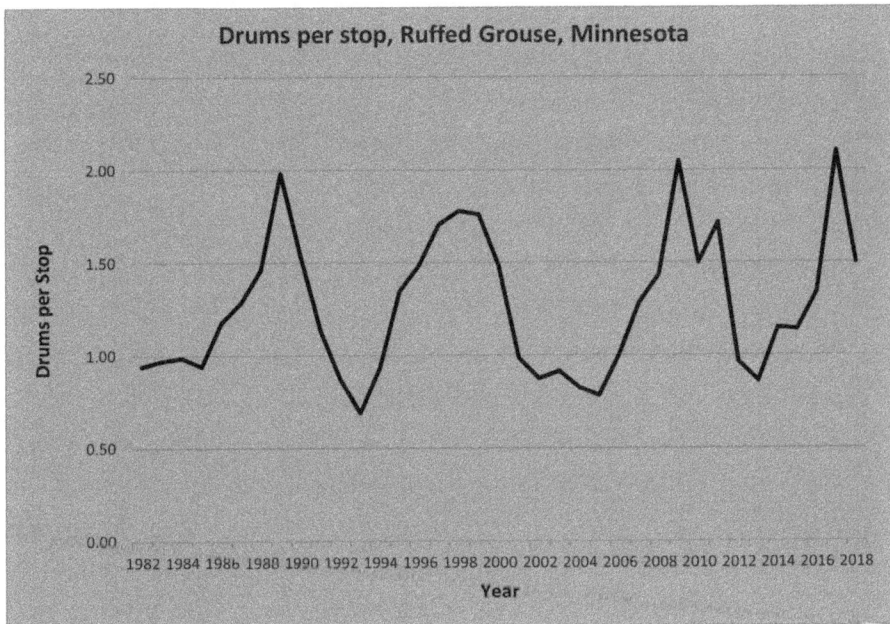

Figure 2-1. Source: Dr. Charlotte Roy, Minnesota Department of Natural Resources (Minnesota Department of Natural Resources).

Plotting the population size of the ruffed grouse over time, shown in Figure 2-1, is a good way to see the effects of limiting factors on a population. The cyclic nature of their population size is a function of the impacts that limiting factors have on the population. When limiting factor impacts decrease, the grouse population grows; when limiting factor impacts increase, the population contracts. "Drums" refers to the sound male ruffed grouse make to attract females, a noise not unlike a drum being played. Ruffed grouse can be hard to see in the woods; biologists use drumming as a proxy to count how many males are in an area. "Drums per stop" refers to the technique biologists use to count grouse: walk on a planned path, pause at specific stops, and listen for drumming. On the assumption that the ratio of males to females is 1:1, overall population size in a given area can be found by doubling the number of drums heard in that area.

Grouse populations are impacted by predators including northern goshawks and great horned owls, fox, fisher, bobcat; and humans hunt ruffed grouse. They may be susceptible to the disease of West Nile virus and to parasites. In Minnesota, they live in habitat of young spruce, birch and aspen forests, where they feed on tree and shrub buds, young leaves and shoots. Their populations tend to fluctuate in ten-year cycles for reasons that are not fully understood but that may relate to predation (Minnesota Department of Natural Resources).

Not all populations of organisms have such regular and distinct peaks and valleys to their population growth cycles. But this graph of the Minnesota grouse population is helpful here to show that populations in the natural world vary based on limiting factor impacts.

Three general observations about the impact of limiting factors are worth noting: first, the larger the population size of an organism, the more prone it is to being heavily impacted by each limiting factor. A large population of necessity requires more of the resources the individuals need to survive which constrains availability of resources for the rest of its population or for additional individuals, thus constraining population growth. In terms of disease, microbes that cause disease typically thrive under the crowded conditions of a large host population: those diseases spread quickly when host individuals are close together, a situation that will limit or decrease the size of the population. In a similar dynamic, predators have an easier time consuming prey when the prey population is large – prey are more visible and susceptible to predation, a situation that will decrease the size of the prey population.

The second general observation: limiting factor pressures change. No environment is completely static: predators migrate into or out of a region, disease patterns change, the food supply will naturally decrease or increase, and an organism's population will respond accordingly to these changes. Any given organism and the limiting factors that affect it all influence each other, and they are all affected by seasonal and long-term weather patterns and dramatic natural events such as flooding, drought, storms, earthquakes, and tornados. When viewed over time, limiting factors define the maximum upper limit of the size of a population of any given organism. Limiting factors will ebb and flow but a region's maximal resource supply and the minimal disease and predation pressure determine the maximum size of a species population.

Third, the limiting factor system brings long-term sustainability to every organism's relationship with their environment. Because population size of every organism is constrained and because limiting factors to a large part are made up of living organisms, populations grow and contract within a matrix of other organisms. The inherent nature of limiting the size of any organism's population through the interactions with other organisms means that an organism can survive over the long-term, effectively forever, or until the environment changes in ways that an organism can't respond to. A balance is reached between all organisms in an ecosystem and is mutually sustaining. Populations are both supported in their growth and constrained from growing unduly large, meaning they can continue in this manner indefinitely. It is at once a simple, elegant, effective and unavoidable system.

The Role of Technology

A fundamental drive in all organisms is simply to survive: individuals seek to harvest and absorb the resources needed to remain healthy, to avoid being prey and to avoid hosting disease. Survival is a foundational drive of all life and limiting factors are effectively elements of constraint working on organisms virtually constantly. Some organisms of the natural world respond to the limiting factors they are constrained by in an interesting way: they create technology to reduce the degree of those constraints.

"Technology" can be defined different ways. Here, "technology" is defined as an object formed by an organism through the manipulation of natural resources, and applied in a way that creates some benefit for the organism. This definition is intentionally broader and less complex than some – the concept of what a technology is, is very simple. In the drive to survive, organisms do all they can to increase their odds of survival and technology does that by reducing the impacts of limiting factors. Technology may refer to a specific constructed object or tool, or it may refer to a collective group of fabricated tools. For example, a wood spear with a stone point is a human technology; and hunting technology includes tools such as the spear, bow and arrow, atlatl and bola.

Many organisms create and use technology: sea-otters use rocks to break the shells of seafood they seek to eat; woodpecker finches use cactus spines to prod otherwise inaccessible corners for food; many birds build nests to reduce predation; Egyptian buzzards break ostrich eggs to consume the contents by dropping stones on them; ravens have been known to drop stones on intruders to defend their nest; leafcutter ants grow fungi for food; herons drop food on a lake surface to draw in fish; elephants use pieces of tree branch to rub or scratch their bodies and to swat away flies; spectacled bears have been observed using sticks to draw food toward them; beaver build lodges and dams for protection and food harvesting; hermit crabs use empty gastropod shells for protection from predation; caddisfly larvae construct cases out of sand, silk and organic matter to provide protection from predators; Howler monkeys in trees have been known to drop branches on intruders below (Beck 1980, 13-45).

What these organisms are doing with their technology is increasing their odds of survival by pushing back against the constraints of limiting factors: improving the odds of resource supply or reducing exposure to disease or predation. Every technology a species produces is unique, a function of their unique physiology, unique biological needs and the raw materials available to them. Their technologies are very simple: based on natural materials found in their environment, secretions from their body and their ability to manipulate the raw materials. An applied technology helps the organism resist the effect of limiting factor pressure so its population is larger than without technology. Limiting factors still limit population size, but to a smaller degree.

Natural Upper Limit of a Population

The impacts of limiting factors on any given species are in constant flux due to the interactions of many environmental variables. Limiting factors in many cases are living organism, meaning they are influenced by their own set of limiting factors; as their populations change, their impacts on other organisms will change. Therefore, limiting factor impacts vary from region to region and they can change over time. The biological world changes constantly, given that it is a function of many organisms interacting with each other influenced by weather, climate and geological forces of the earth. Very little in the living world remains static for long.

The net effect of limiting factors on an organism determines the population size of that organism at any moment. If you tracked the population size of an organism over a period of time and plotted this data on a graph, the population changes would be evident as ups and downs, with peaks and troughs. With this graph, you could see that one of the peaks, the tallest, represents the largest size of the organism's population over the period. This maximum population size is where the supportive aspects of limiting factors were greatest and the constraining aspects of the limiting factors were minimized. This point of maximum population size is the natural upper limit of the population over that time period.

Limiting factors define the natural upper limit of an organism's population; in being limiting, they constrain population growth. Graphically, they exert pressure down on the population. We can see this below on the graph showing the ruffed grouse population in Minnesota.

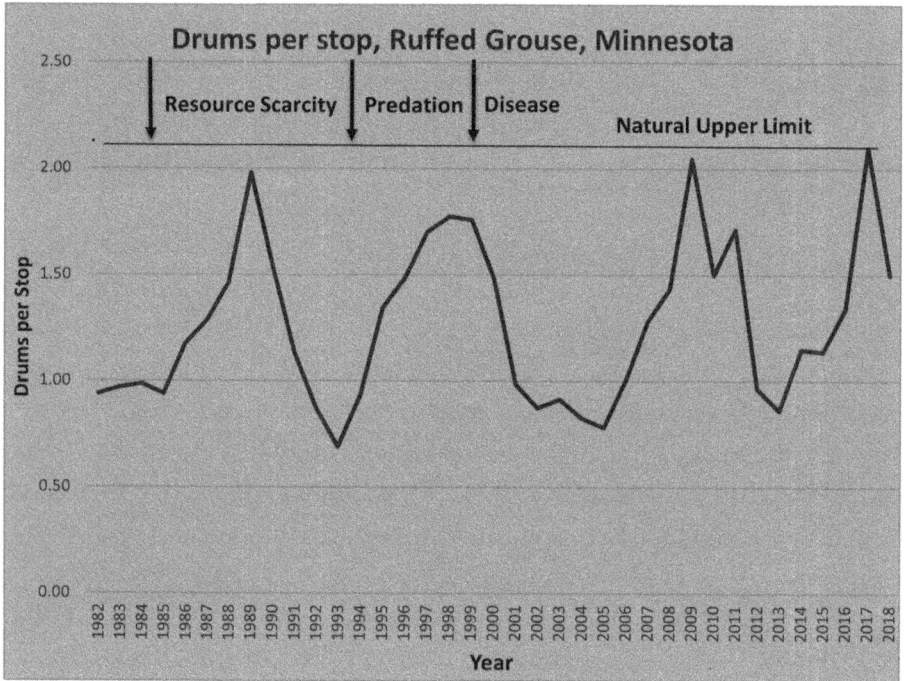

Figure 2-2. Source of ruffed grouse data: Dr. Charlotte Roy, Minnesota Department of Natural Resources

Graph 2-2 shows the natural upper limit of the grouse population for the time period 1982 – 2018 being 2.1 drums per stop in 2017. The horizontal line touching that maximum population size represents the natural upper limit over this time period. The three arrows pointing down give a visual representation of the constraint limiting factors exert on the population size.

Human Hunter/gatherers

Archaeological evidence in Africa indicates that *Homo sapiens* evolved from *Homo erectus* by about 315,000 BCE (Hublin 2017, 290). Humans dispersed from Africa to distant parts of the earth, adapting to different ecological niches. As hunter/gatherers, regional populations probably followed the same pattern as other organisms, increasing and decreasing in size in response to food supply, disease and predation (Coale 1974, 47). Over the long-term our population grew slowly (Kremer 1993, 683).

The specifics of what our hunter/gatherer ancestors ate, their behavior and how they related to each other were unique to our species, just as the natural history of every species is unique. Hunter/gatherers maintained small clan populations of 50 people or fewer by constraining reproduction or by splitting clans into separate groups and dispersing to avoid overconsumption of local resources and to manage internal clan dynamics. Hunter/gatherers were nomadic, moving in response to the dynamics of their environment and not over-consuming resources. They had few possessions and used little technology. They maintained deep and broad knowledge about their regional ecosystem, allowing them to find and harvest what they needed to survive. One thread of anthropology believes hunter/gatherers maintained comfortable, low-stress lives, capable of harvesting what they needed to survive (Sahlins 1972, 1-39; Gowdy 1999, 391; Cohen 1977, 20).

Every regional hunter/gatherer population maintained a direct, intimate relationship with their environment, the source of the resources they needed to survive. Hunter/gatherers were deeply skilled in finding what they needed to survive: they knew the natural history of the different plants and animals in their region, knew where to access water, were able to find resources to create a shelter or craft a technology when needed. They ate a varied diet and apparently contracted few diseases. It is believed they lived secure, comfortable lives (Gowdy 1999, 397; Ammerman 1975, 220).

Just as important as hunting and gathering the physical resources they needed to live was creating a community that generated a healthy culture. Failure to do so could easily result in fractures within the group, threatening individual and clan survival. To maintain group coherence they managed their cultures to generate harmony and stability. Clans had no class hierarchy, political structure or central leadership. Personal conflicts were brought into the open for the clan to help resolve. Inequality within the clan was minimized through policies of sharing and reciprocity. Even in larger populations with head-men or leaders, ceremonies of gift-giving to the leaders were expected to be followed at some point by giving the gifts away (Lee 1999, 3-4; Ember 2014, 3; Sahlins 1972, 189-194).

By generating close, personal relationships within the clan, with no power structure and an impetus to support each other and maintain harmony, hunter/gatherer culture was simple. Inter-clan dynamics were straightforward and balanced. They sought to maintain equanimity with each other and with their environment.

The economies of hunter/gatherers were simple. Theirs were pure free market economies based on the microeconomics of supply and demand. When the population demanded a resource they looked to the environment to supply it. Clans understood that they needed to consume prudently; by harvesting some but not all of a resource the resource could regenerate and be available in the future. In ecological terms this is a sustainable supply-and-demand relationship that could go on indefinitely. The aspect of sustainability is a foundational aspect of biological life systems.

While individual clans did not grow large, the global human population grew over the long-term, from possibly just a few thousand early in our evolution to between five and 10 million in 10,500 BCE (Coale 1974, 41). Prior to about 10,500 years ago our ancestors were one of many species living in a sustainable relationship with their environment in a balanced and flexible equilibrium with all other species around them (Ammerman 1975, 220-221). This dynamic changed when humans engaged in what is now known as the Neolithic Revolution.

Chapter 3

A Transition

Between about 10,500 and 4,500 years ago, up to nine independent, regional populations of humans around the globe put down roots. They turned from a hunter/gatherer lifestyle to a lifestyle of agriculture at what is called the Neolithic Revolution (Diamond 2003, 597). There is tremendous discussion in the anthropological community about why humans morphed from being hunter/gatherers to agriculturalists. There are many, wide-ranging theories, with no definite consensus. V. Gordon Childe proposed the 'oasis theory' in 1936 (Childe 1951, 59) based on wide-spread changes in climate; Robert Braidwood developed his 'hilly flanks' theory, that the Neolithic transition could only emerge in areas that inherently maintained specific conditions suitable for human domestication of plants and animals (Braidwood 1964, 91-97). There are several other hypotheses by other researchers (Weisdorf 2005, 561-568; Redman 1978, 89-103; Gebauer 1992, 1-3). Price and Gebauer note that the many theories of why humans turned to agriculture tend to come to three factors affecting humans at the time of the Neolithic: climatic or environmental change, population pressure and changes in social organization (Price 1995, 4). Mark Nathan Cohen assessed human demographic dynamics of hunter/gatherers and wrote that a large population was the driving force for the shift to agriculture (Cohen 1977, 11-13). Cohen's key ideas are important and will continue to show up in this book.

The shift to agriculture was not quick or linear. Between the hunter/gatherer and the agricultural systems intermediate types are recognized. Economist Ester Boserup identified six different systems: gathering (the most extensive and the least intensive system), forest-fallow, bush-fallow, short-fallow, annual cropping and multicropping (the most intensive and the least extensive system) (Boserup 1981, 18-20). In each system humans applied different techniques, to varying degrees, as they saw fit; they scattered seeds, cleared land, applied fertilizers and were semi-sedentary. The different systems are not rigidly defined – lifestyle flexibility allows populations to incorporate specific aspects of food harvesting systems based on what worked for them. Some hunter/gatherer populations had been informally planting seeds, clearing land, and practicing plant and animal domestication on a small scale well before the Neolithic. And there is evidence that in some regions populations went back and forth between hunter/gatherer and agriculture or one of the other systems over a period of time before becoming fully agricultural (Borrell 2015, 1-19).

The Neolithic was an enormously significant transition for the human species. Pre-Neolithic humans were hunter/gatherers, fitting into the ecosystems they occupied along with all other species, applying their biological traits to the business of survival just as all other organisms hunt and gather with the intent to survive. At the Neolithic we changed from being hunter/gatherers to being agriculturalists. After hundreds of thousands of years of living a lifestyle consistent with how biological systems functioned our species made a change to a new system in nine independent regional populations around the globe.

Characteristics of an Agricultural Lifestyle

In shifting to agriculture, the characteristics of our lifestyle changed. These changes were not simply a continuation of hunter/gatherer characteristics but significant changes in how we structured our means of survival.

Agriculture

The shift to agriculture brought systemic changes to the way humans gained the resource of food. Instead of accepting what the environment supplied we began to proactively grow the food we needed. Agriculture consists of four broad elements: plant cultivation, harvesting crops, storage of crops, and control of plant and animal breeding for the desired traits (Redman 1878, 91). We began to collect plant seeds, clear fields, plant crops, augment the land with fertilizers and other additions, irrigate where rain was uncertain, terrace in steep areas, protect crops from pests, harvest crops, develop ways to store harvested crops, domesticate plants to improve productivity and domesticate animals for their ability to work and as food.

An advantage of agriculture is that it is a scalable way to supply our food. An increase in demand could be met with increased supply by expanding existing fields, augmenting soils with fertilizers, irrigating crops and breeding more productive plants.

Settlements and Cities

At the Neolithic people stopped being nomadic – nomadism is not feasible when a population develops a plot of land for long-term benefits. Improving the land, planting, harvesting and storing the crop calls for being sedentary. With sedentism people congregated in settlements larger than 50 people and built permanent structures for living and storing the food supply. As regional populations grew, people grouped together into larger settlements, no longer constrained in size by the limits of the naturally occurring food supply. Over time settlements grew larger and became cities. In the Sumer region in ancient Mesopotamia the first cities of several thousand people formed about 3000 BCE (McNeill 2003, 45).

Technology

Along with agriculture, people created more technology: plows, sickles and hoes for working the land, pottery and buildings for food storage, grinding tools for processing grain, crop and animal domestication and stone axes and adzes. In some regions the creation of the wheel facilitated moving products. Evolving human technology was possible when people started being sedentary: nomadism implied carrying all they owned, reducing the impetus to create and acquire items. Evolving technology allowed people to create tools to increase productivity of their fields, helpful to support a growing population.

Government

Growing populations and the formation of settlements and cities prompted the formation of governments with central leadership. Different populations structured their governments differently but each vested decision-making and policy-setting power to their leaders. With government, the population became politically structured (Kent 1989, 708-709; Naroll 1956, 687-696; Carneiro 1970).

Economics

The economy of a sedentary agricultural population differs significantly from the economy of nomadic hunter/gatherers. Agriculture implies active management of the environment: making changes to the land to be able to grow crops and to increase productivity. With land management, people have a vested interest in fields they have spent energy to improve and with those investments come land ownership, a foundational concept in complex economies but entirely foreign to hunter/gatherers. With the capacity to increase productivity and profit owners seek to create more technology that increases efficiency and productivity. Land ownership implies economic inequality: a few could own property and accrue the benefits while many could not.

With evolved technology and increased productivity the economy grew and with that growth came economic complexity. Over time, new elements emerged in the economy: hired labor, specialization, surplus production, investment and risk, future value, credit, profit, income, special interests, savings and wealth. Trade with distant populations increased prompting management through tariffs and regulations (Samuelson 2001, 25-35). With a large population and a complex economy, hierarchy develops. Wealth and power accrue to some more than others. Inequality is inevitable.

All these characteristics of a complex and growing economy are foreign to hunter/gatherers. A complex agricultural economy is no longer a pure free-market economy. Microeconomic analysis is still valuable, but macroeconomic theory is necessary to attempt to understand a complex economy.

Complexity

An interesting population dynamic that emerged at the Neolithic was the development of complex economies and complex cultures. Complexity is a significant theme of study for economists and anthropologists and each field tends to look at complexity in slightly different ways. For this book, a simple definition is important, to be clear about the concept and to relate to the perspectives of different fields. Here, complexity refers to what results when a system acquires more parts and there are more interactions between those parts (Price 1985, 7). A simple economy has few transactions and few types of transactions between those who supply and those who demand resources. A complex economy has many players, each of whom are involved in supply and demand transactions.

Hunter/gatherer clans and cultures were relatively simple, with a low level of technology, a supply and demand economy based solely on microeconomic transactions and a culture contained within a small population of 50 individuals or fewer. Neolithic dynamics generated complexity. More and evolving technologies meant we could manage and actively produce more items, leading to a larger and growing economy and a larger and growing population. There were more interacting parts in Neolithic populations.

Cultural differences between simple hunter/gatherers and complex agriculturalists are not strict. Anthropologists recognize that there have been hunter/gatherer populations with complex characteristics, especially in regions with rich, yearlong food sources (Price 1985, 3-8). And in some cases, complex hunter/gatherer cultures may have been in transition to an agricultural lifestyle (Belfer-Cohen 1991, 171-173; Henry 1985, 365-366).

Inequality

With the development of agriculture and larger populations, the political structure, the larger economy and the complexity that develops, inequality between individuals of the population emerges. Where a small clan of hunter/gatherers can manage their group dynamics to maintain equality, the larger and more complex dynamics allow inequality to be established.

This is an enormously significant change. Hunter/gatherers structured their culture to be harmonious and stable for the long-term, and one of the ways they did that was to maintain equality among individuals of the clan. Equal treatment within the clan generates an atmosphere of support for everyone. Resources distributed with reciprocity and equality leads to mutual support, and both individuals and the clan have better odds of survival.

With agriculture comes a growing population, a more complex culture, a larger and more complex economy. These changes bring hierarchy to the population, with some people performing specialized tasks, some individuals owning the means of production, and someone making macro-scale decisions for the population. With hierarchy, some individuals gain more power or resources; consequently, inequality emerges. Whether sanctioned by the group, such as an elected leader, or implicitly recognized, such as a landowner, individuals of the group are no longer considered to be equal. Resource allocation has changed and the power dynamic within the population is no longer one of equality. Without equality the culture is inherently less stable.

Agriculture and Population Growth

Another characteristic of agriculture is significant enough to warrant its own heading. Hunter/gatherers maintained small clan sizes to manage stable and harmonious relationships within the clan and to prevent overconsuming resources from the environment. Regional populations of every organism have a natural upper limit based on limiting factors of the region, and human hunter/gatherers were no exception. Just before shifting to agriculture at the Neolithic each of the nine regional populations around the globe was likely very close to their natural upper limit, a function of very slow population growth over tens of thousands of years and their impetus to keep clans small.

At the Neolithic the shift to agriculture meant more than growing food as a substitute to hunting and gathering it. Shifting from being hunter/gatherers to being agriculturalists meant our population no longer relied on what the environment provided naturally for our food supply. By actively growing what they needed humans lessened the impact of the limiting factor of resource supply. With a diminished limiting factor the human population could grow. And because agriculture could expand in acreage and increase in productivity, the amount of food grown could increase as the population increased. This population growth was something new – it put our species on a new path. We were still constrained by limiting factors but agriculture meant we were constrained to a lesser degree. Our population grew beyond the natural upper limit and kept growing. This was a unique spot for the human species.

Population growth drove many of the characteristics of the Neolithic populations noted above. A large and growing population prompted a desire for increased productivity of the fields, stimulating technology evolution. More technology and more productivity prompts economic growth and complexity and increased trade outside the local population. A large population inevitably generates political structure to manage the more complex dynamics of the population.

This is a fascinating dynamic: when a population constraint inherent in hunter/gatherer populations was removed, our population grew, allowing new and different elements to emerge. With this dynamic our species took a different path, unique in the natural world.

Transition Summary

Hunter/gatherers lived comfortable lives, readily harvesting the resources they required to live and maintaining healthy and stable relationships within their clans. Their awareness of environmental constraints and living in a balanced relationship with their environment in part resulted in little technological development and an extremely slow population growth. Whether by conscious choice or by the ruthless constraints of limited supply, hunter/gatherer clans maintained small populations, not growing beyond what their immediate environment could support.

With the Neolithic Revolution we shifted to a significantly different lifestyle. The transition happened slowly over time and began at different times in different regions. Each population incorporated regional plants and animals into their agriculture based on the local species suitable for domestication and each region created a unique political structure, but all nine regions developed agriculture and developed similar population characteristics. Technology evolved, political structure developed, their economies and trade expanded, and their populations grew.

> In 2003 I was a member of a canoe trip on a river in the wild, northern reaches of Canada. We were flown in to a remote lake, paddled for five weeks and finished at an Inuit settlement close to Hudson's Bay. The group spent several days there before flying out, and we got to know some of the inhabitants, one of whom was an Inuit man more or less 90 years old. Barnabus was wonderful – quiet, open, and he knew enough English for us to talk. Given his age, and knowing a little of the history of the area, I realized that he must have spent his childhood in a completely undeveloped

region of northern Canada ("on the land," as they called living removed from any development), before moving to a settlement formed by the government. This meant Barnabus had effectively lived in the Stone Age for the first part of his life.

I had to ask, "Barnabus, how was growing up on the land compared to living in town?"

From his body language, from the look in his eyes and from the brevity of his answer, I knew he understood precisely what I was asking. He said, "Living in town is easier. Living on the land is better."

A simple story, but it struck me as profound. Barnabus straddled both worlds. He knew what it was like to live as a hunter/gatherer; he knew life in the developed world; and he could discern the difference between the benefits of an easier life with running water, indoor toilets and purchased food and the quality of life that came from living close to the environment that nourished him. When I talked with other Inuit about their extended trips out into the wilds surrounding their town, they never spoke about any hardship of living on the land – they glowed when they described using their knowledge of the land to hunt and camping in a tent with no one else in sight. Living on the land was vital to their spirits.

Chapter 4

Breaching the Natural Upper Limit

A population existing comfortably below the natural upper limit feels very little impetus to change from its mode of hunting and gathering the resources needed to survive. If people are harvesting what they need with little effort or stress there is no pressure to effect change. *Status quo* is the rational option. Anthropologists tell us hunter/gatherers were highly skilled, competent and comfortable at what they do. They knew how to survive in their environment (Binford 1983, 204-208). That they did so for thousands of years testifies to their competence (Gowdy 1999, 391).

Populations of organisms have options when they feel resource availability stress. They can harvest resources less desirable than optimal; they can harvest optimal resources more intensively than they would normally; they can move locations to a place of resource abundance; they can divide their population into smaller groups to spread resource consumption; they can actively constrain the size of their population by limiting reproduction; or they can create technology that lets them access resources not available otherwise (Cohen 1989, 20-22).

However, a population that has been close to the natural upper limit for a long time may not have all those options available to it. If people have already harvested the desired resources intensively, if they have already harvested the less desirable resource options and if their region's habitable locations are inaccessible to them, they need to constrain or diminish the size of their population or find some way not to be constrained by the limiting factor of resource scarcity. If they do not decrease their draw on the environment or access necessary resources nature will decrease the size of their population through starvation and exposure.

Two elements were necessary to prompt the human population to go beyond the natural upper limit imposed by limiting factors: a population close to their natural upper limit and therefore feeling the effects of resource scarcity pressure and an ability to resist the impacts of limiting factor pressure to more than a small degree. To resist limiting factor pressure humans needed to be able to create technology.

Humans are the only organism with an ability to create complex technology. Our uniquely evolved physiology, including large brains, opposable thumbs and an ability to communicate within our species allows us to problem-solve, experiment, create and disperse technology that allows us to push back the impact of limiting factors that then allows our population to grow above the natural upper limit. Other organisms create simple technology – caddis-fly cases, beaver lodges and dams, leaf-cutter ants' agriculture – with basic natural resources, but we can create technology that is more elaborate, more complex, and more effective at reducing the impacts of limiting factors.

The changes that occurred starting 10,500 years ago in the regional human populations allowed them to grow more food than was available in their natural environments. Agriculture was a way to push back against the constraint of limited resource supply. Population growth was slow but it expanded above the size imposed by the natural, undeveloped environment. Uniquely in the history of the earth, those regional populations grew larger than the natural upper limit imposed by biological limiting factors.

Chapter 5

Dynamics of the Natural Upper Limit

To propose that regional populations of humans just before the Neolithic were close to their natural upper limit begs some questions: without accurate long-term data of the sizes of regional populations, how can we know they were close to the natural upper limit? Even if we did have that data, how can we know what a natural upper limit is for a population of humans? Can we know they were feeling resource scarcity at the Neolithic?

In a similar vein, estimates of the maximum number of people who could live on the earth are numerous. The range is enormous, from 500 million, to tens of billions, to a surreal one trillion people, and even higher. All are based on different research assumptions: the future availability of water and food, how we can continue increasing food production and how much more food we can grow, topsoil formation and depletion, the nitrogen and phosphorus cycle in agriculture, fertility rates of humans at different points in the population growth curve, medical technology advancement, general technology advancement, and more (Cohen 1995, 402-425).

From a biological perspective a single number does not help because environmental variables that change constantly influence population size. Seeking an estimate of the maximum number of people the earth can support begs the larger questions: do we really need to know this? Does this question, or any answers to it, really help us?

The question of discerning the size of a human population close to the natural upper limit, or being able to know when a population feels resource scarcity, and the question of the maximum number of people the earth can support, are all different questions, but any helpful answer to each of them comes to the same point. The natural upper limit and the maximum number of people the earth or a region can support are variable, based on two factors: first, the variable impact of limiting factors needs to be recognized; they change as the environment changes. Second, the questions concerning the maximum size of the human population need to be qualified: maximum population size over the short- or the long-term. Over long periods, maximum population size will increase and decrease following long-term environmental trends: glaciation, geologic changes, long-term patterns of rainfall and plant growth. In the short-term, the maximum will be what environmental limiting factors define at that moment, including variables of rainfall, severe storm occurrence, geologic traumas such as earthquake and volcanos and patterns of disease. As well, the assessment of maximum population in the short-term needs to qualify what level of technology is assumed. Short-term population numbers can increase above the sustainable level when technology is applied. If no technologies were used to push back limiting factor pressure, the population would be smaller than if technology is applied.

The question of a maximum population size can be modified to focus on what would help. What if we ask: "How many people can the earth support sustainably?" The qualifying word refers to the ecological meaning: a population that can access the resources it needs to survive from an environment that can supply those resources over the long-term, effectively forever, is a sustainable relationship. Modified in this way, the question of the maximum number of people the earth can support is similar to the question: "what is the population at the natural upper limit?" Other than humans, every organism on earth has been and is in a sustainable relationship with its environment. A sustainable population will fluctuate to be smaller and larger over time, but will remain extant because enough individuals gain the resources that allow the species to survive on a long-term basis. Any helpful concept of the largest number of people the earth can support has to factor in the biological elements of people harvesting and of the environment providing, sustainably, forever. This biological, sustainable qualification implies that there is no single, correct numeric answer.

For understanding the population growth dynamics of the human species in the context of this book, the concept of a natural upper limit is important. Our discussion of the Neolithic Revolution and consequent changes in human culture were based in part on the assumption that each of the independent populations of humans were at their natural upper limit just prior to their Neolithic transitions, and this implies a knowable, defined number. This is a somewhat presumptuous proposition: we have no accurate data on the number of people alive at points of Neolithic transition, and even if we did, a specific number would not explain the biological dynamics that the populations were experiencing. A specific size of a population is not as important as the resource stress it feels.

Every organism needs to harvest resources to survive. When needed resources are scarce, survival is in question and individuals and the population feel stress. Resource scarcity stress occurs when the population is close to the natural upper limit. Scarcity is a relative value. Biologically, a population does not feel resource stress at the point when a needed resource is completely consumed, but at some point well before that. When individuals of a species spend more energy seeking what they need than they used to, or that feels acceptable, that resource is scarce. A regional population of squirrels will not feel food stress at the moment they consume the last acorn, but before that point, when acorns are hard to find. Humans were skilled hunter/gatherers for thousands of years, and from what we can discern they lived comfortable lives. With little stress, there was no reason to change lifestyle.

It is significant that at the Neolithic, nine separate human populations shifted to a lifestyle that provided a less nutritious diet, was more prone to disease (Cohen 1989, 132-141) and was more work than being hunter/gatherers (Cohen 1977, 30; Sahlins 1972, 14-18). These shifts from being hunter/gatherers to agriculturalists imply the populations were feeling resource stress —this was not a chance direction, or a chance decision by a single population. The realm that concerned them directed the change in their resource supply systems. They would have no other reason to change their system of food gathering. If predators or disease decimated their populations, they would likely not have opted to shift to agriculture unless they knew such a change would keep away predators or illness. From the distance of thousands of years, we cannot know precisely what pre-Neolithic populations were experiencing, but that they made a change from hunter/ gatherer to agriculture implies they were at the natural upper limit, due to resource scarcity.

Unique in the natural world humans had the ability, untested before the Neolithic, to develop complex technology. Any other organism feeling resource scarcity would have been unable to create agriculture and complex technology for a lack of ability to do so, meaning that their population would have had to stabilize or shrink —the natural world would not allow otherwise. Expanding into our awareness of being capable of manipulating resources to fabricate what was needed to survive, humans developed agriculture.

A separate perspective on the question of where the natural upper limit of the population lies concerns the impact people have on the environment. Limiting factors inherently preclude a population from generating widespread pollution and resource depletion. If an organism creates pollution that is detrimental to its health, the population would decrease. If an organism consumes resources faster than the natural rate of regeneration, the population would decrease. That humans generate long-term and widespread pollution and are depleting resources faster than the natural ecosystem can replenish them implies *Homo sapiens* is beyond the natural upper limit.

Chapter 6

Cost/Benefit of Neolithic Changes

Like all other organisms in the world pre-Neolithic humans worked with the supply and the scarcity of the environment to hunt and gather what they needed to survive – a system that worked for hundreds of thousands of years. At the point of the Neolithic, we opted to grow our food, producing more on a specific plot than the environment could supply naturally, therefore reducing some of the constraint of the limiting factor of resource supply and allowing our population to grow beyond the natural upper limit. Our large population and applied technology inevitably incurred complex population dynamics not present in hunter/gatherer populations. We found ourselves in a unique situation in the natural world. A population above the natural upper limit did not simply mean 'more people'. The characteristics of our population and the population dynamics that emerged were unique, unlike those of hunter/gatherers. At the Neolithic, we embraced a different lifestyle, including aspects of sedentism, forming settlements and eventually cities, growing our population and our economy. There were both costs and benefits to this lifestyle.

Costs of Agriculture

Creating and applying agricultural technology to grow our food supply brought a handful of costs.

Losing Close Relationships with Each Other

Hunter/gatherers structured their clans to be culturally harmonious and stable: a small clan size; sharing and reciprocity of resources; no hierarchy or central leadership; public resolution of disagreements. These traits all worked to keep relationships within the clan close and harmonious, to keep the clan working together to survive.

The shift to large, hierarchical and structured populations weakened or negated the close, personal relationships of clans. The emergence of political structure and a complex economy created hierarchy and classes, and generated inequality within the population. In a larger population, close relationships between each individual were difficult to maintain – there were simply too many people for others to know personally. Small communities could form within large ones, but the reality of large populations is that individuals were not close to many of the people around them (Wirth 1938, 10-14; Simmel 1902, 2-4, 12).

While the hunter/gatherer culture maintained harmony and equality within the clan, in a large population, less harmonious behaviors can emerge. The anonymity and personal disconnection in a large population allows some less hospitable characteristics of our human nature to be expressed.

Losing Close Relationship with Their Environment

Hunter/gatherers have a close and direct connection with the land they inhabit. They accept what the environment supplies with little manipulation, drawing directly on the resources around them: food, water, resources for shelter and clothing, fuel for fire. Every regional hunter/gatherer population maintains an intimate knowledge of the natural history of the plants and animals in their region. Knowledge is gained, passed down, evolves as the environment changes and is needed to survive as a hunter/gatherer.

As agriculturalists, humans maintained less awareness of their ecosystem as they gained awareness of how to manipulate land and resources to produce what they wanted. People no longer relied on an undeveloped, natural land to survive or on maintaining a stable supply and demand relationship with their environment. In committing to technology people relied on their ability to develop land: to create fields, create technologies, to grow crops that would provide sustenance for the population. In growing crops, people learned a lot about specific plants, but lost broad knowledge about the many other plants in their region. In domesticating animals for food and labor they lost intimacy with animals they used to hunt. Agriculture and the use of technology broke the thread of knowledge people maintain when their lives are based on harvesting resources from the ecosystem. We shifted away from a relationship with the environment based on respect, to seeing the environment as a source we could take from without regard for consequences (Redman 1999, 25; Agnoletti 2014, vi).

Hunter/gatherers used muscle energy to do all physical labor; with agriculture, humans created technologies to do work in our place. When external energy sources, such as wind and flowing water, were harnessed to power grain mills, saw wood, pump water and power machinery, people became less engaged with the labor and personal energy required to provide them with sustenance. When less energy is expended by people, they're less connected with the work of transporting themselves, digging holes in the ground, planting seeds and harvesting crops, moving products, carrying water, cutting wood and many other tasks that were previously performed by people.

Diet

The Neolithic diet was less healthy than the diet of hunter/gatherers. Agriculturalists tend to grow one or a few crops intensively, offering little variety of diet, while hunter/gatherers selected from a broad range of foods available to them, and the variety provided a healthy diet. Relative to hunter/gatherers, the health of early generations of agriculturalists declined: body size, body height and bone thickness decreased at the Neolithic (Livi-Bacci 2017, 38; Cohen 1989, 132-134).

Open to Failure

Another cost of our evolved technology is that agriculture is more open to failure than a natural ecosystem and requires human manipulation and vigilance to guard against crop loss (Cohen 1989, 133-135). The natural biological systems hunter/gatherers work with tend to be diverse and resilient to weather events and nature does the work of storing, planting and growing all the resources needed to live. Agricultural systems are not so resilient. A weather anomaly – extreme heat, cold or wind, drought, flooding – or a biological event of insects or disease can decimate agricultural crops in a field or region or ruin stored crops. Monoculture is inherently susceptible to decimation by disease, insects, or weather events; simply by being unnaturally concentrated in one location, a species is vulnerable to decimation.

Many advanced populations in the past have collapsed when their culture, built up on agriculture and technology, could not withstand a shock. Different authors define societal collapse differently. Glenn M. Schwartz offers a good description: "In the archaeological literature, collapse usually entails some or all of the following: the fragmentation of states into smaller political entities; the partial abandonment or complete desertion of urban centers, along with the loss or depletion of their centralizing functions; the breakdown of regional economic systems; and the failure of civilizational ideologies" (Schwartz 2006, 5-6).

Railey and Reycraft give an excellent discussion about the approaches different anthropologists have taken in assessing the reasons for societal collapses in the past (Railey 2008, 4-12). Historical collapses of cultures have occurred for different reasons, caused by factors either internal or external to the population. But the fundamental point is that a collapsed culture was overextended: the population had grown large using agriculture and technology and was dependent on those to maintain its size and strength. When a shock occurs — drought, disease, war, internal dissention, political shift — the overextended culture could not maintain its complexity and collapsed.

Resilience to collapse is precisely what hunter/gatherers and naturally occurring biomes maintain. Disease occurs, but it can rarely spread far or fast. An anomalous weather event can happen, but rarely with an impact of decimating a species for more than a short period. Flexibility and resilience in growth and consumption are inherent in natural biological systems and in the resource supply and demand economy adhered to by hunter/gatherers.

Personal Effort

Anthropologists have discerned that hunter/gatherers spend fewer hours per person per day harvesting the food they need than agriculturalists spend in growing and harvesting their crops (Gowdy 1999, 392-393). It simply takes more time and personal effort to grow crops than it does to hunt and gather the resources necessary to support people. Anthropological research finds that most hunter/gatherers were capable of harvesting what they needed, without pressure. It takes more labor to perform all the steps of agriculture than it does to harvest what is available from the environment.

Contact with Disease

The shift to agriculture brought people into close contact with many diseases that hunter/gatherers did not interact with (Cohen 1989, 135-138). Several factors were responsible for this. By being sedentary and living in permanent structures, people were living in close proximity to higher concentrations of human and domesticated animal wastes, and possibly contaminating their water supply with disease microbes. And rodents and insects were attracted to the permanent structures people constructed for housing and food storage; humans came in contact with the microbes those organisms carried, including diseases (Barnes 2005, 51-52).

As well, when humans domesticated animals, they lived in close contact with those animals, some of whom carried organisms that were disease causing to humans. The diseases of rabies, measles, toxoplasmosis, influenza, trichinosis, brucellosis, anthrax tuberculosis, roundworms and tapeworms all resulted from human contact with diseases originating in domesticated animals.

A complicating factor to the incidence of disease among agriculturalists is that people changed to a lifestyle that aids in disease transmission. Many disease organisms thrive in crowded conditions – disease microbes spread more readily with close contact between host and victims, conditions that occur in settlements with large and dense populations. As well, after the Neolithic trade between settlements expanded. Along with the goods we shipped to distant markets we transported disease organisms, allowing them to populate new regions. Epidemics are a function of a readily transmitted disease meeting a densely pack population. In the mid-1300s, shipping products between population centers along trade routes by land and water also transported rats that carried fleas infected with *Yersinia pestis*, causing the Black Death epidemic in Asia and Europe.

Environmental Degradation

Environmental degradation is a side effect of complex technology and a population larger than the natural upper limit. By-products from the creation and application of technologies make their way into the soils, waters and atmosphere of the planet. Hunter/gatherers' simple technologies did not generate environmental degradation, either in their manufacture or use, or after they were discarded, but starting at the Neolithic, technology production and use resulted in toxic by-products.

The human population, grown large from the application of technology, requires more resources to support people: food, water, resources to fabricate shelter, clothing, energy to heat homes and businesses and energy to transport people and products. While hunter/gatherers consumed resources at rates lower than ecosystems could regenerate them, regional populations after the Neolithic were larger than the natural upper limit and therefore required more resources than the earth could supply naturally. The result of this forced resource production is resource depletion.

Shifting to agriculture at the Neolithic probably did not result in changes that caused immediate environmental degradation. But as technology evolved and the population grew, environmental impacts increased. Two of the environmental impacts that resulted from the change in lifestyle include salinization of agricultural land from irrigation and pollution of water and soil from concentrated human and animal wastes. Later, with the creation of advanced and more complex technologies, pollution would be a significant problem.

Trapped by Technology

When a population accepts a technology, limiting factor pressure is reduced, allowing the population to increase to a new ceiling. In this way, agriculture and technology have the effect of propping up the population at an inflated level relative to a population without technology. Applied technology has the effect of 'improving the quality of life' of the population, but along with that positive aspect is the downside that if the supporting technology is removed for any reason, the population is unsupported and will collapse (McNeill 2003, 213-214). The population size will decrease to the point it is supported by the environment and any other applied technology. Technology is fallible; the risk is acute.

A related problem is that a population grown larger than the natural upper limit is less flexible in its options if a significant stress emerges. Hunter/gatherers have several options in the face of environmental stress: move to a different location where conditions are better, harvest less than optimal resources if optimal are not available, harvest needed resources more intensively, and split the clan into two groups to decrease localized consumption pressure (Binford 1983, 208-211). However, agriculturalists do not have that flexibility: tied to their fields, shelters and technology, they cannot move or quickly change consumption patterns. They are reliant on the systems they have set up for support: sizable harvests from their fields, technology to process and store those harvests, a supply of water and perhaps goods from trade. If something changes that brings a stress to the population – a poor harvest, decreased water supply, political dysfunction, internal strife, and attack from an outside group – their ability to adjust is limited.

Benefit of Agriculture

Given the sizable list of downsides, why would hunter/gatherers opt to change to an agricultural lifestyle? Without knowing all the factors affecting individual populations over 10,000 years ago and assuming that large populations were pushing the capacity of their environments to support them, it seems likely the primary motivator to shift to agriculture was that it offered a path of survival without decreasing their population size. It's possible that in a given region, if hunter/gatherer clans were surrounded by other clans, each may have felt pressure to keep a large population or risk being outcompeted or overrun by neighbors – a larger population has more physical and emotional clout than a smaller population. This pressure to maintain and grow a population would have been even more acute if neighboring clans had a history of violence and warfare. Survival in a sparsely populated region implied living within the bounds of natural environmental supply. But surviving in a region with hungry and violent neighbors may have led to a different path: a large population and additional technology. The costs of agriculture are numerous, but none weighs as heavily as the cost of failure to survive. The shift to agriculture was beneficial for one reason.

Reduced Limiting Factor Pressure

By embracing agricultural technology, humans decreased the impact of limited resource supply on their population. Agricultural technology effectively pushed up the natural upper limit of their population. Their agriculturally-based diet was not as healthy as hunter/gatherer's, but they were able to simply grow more calories, support their population, and grow their population; these were all significant benefits to them. Agriculture was a means to survive.

One reason agriculture was effective to this end is that it is scalable – it can expand as the population grows. This meant that not only could agriculture support the large population as it was transitioning away from hunting and gathering; agriculture and related technology could expand as the population kept growing. Not only did agricultural technology push up the natural upper limit of the human population, technology kept evolving as the population grew, continually resetting the upper maximum size of the population.

In the 10,500 years since the Neolithic Revolution our technology has evolved significantly. It now reduces the impact of all limiting factors, and goes further to simply make our lives easier on many fronts. Agricultural technology has continued to increase in productivity, and medical technology has dramatically reduced the incidence of disease on our population. We are not highly impacted by predation, although the technology of firearms helps reduce that impact, where present. Beyond reducing the impacts of limiting factors much of our long history of technology – shelter technology, transportation technology, electric motors that power tools to do work in place of muscle power – simply makes our lives easier and allows for less personal energy expenditure to perform the business of surviving, which allows us to expend energy in reproducing.

> While shoveling after a recent snowfall, I walked down my sidewalk and chatted with Aaron, a neighbor of mine. It is nice to have a reason go outside and talk with neighbors, since we tend to be more house-bound in the winter. Aaron and I got to talking about snowblowers, technology, and its effect on people, and he made an observation that struck me as directly on-target. He looked both bemused and puzzled when he observed that we crave the ease that technology brings us, but we are worse-off for it, and turn into blobs when we do not engage in physical work. Aaron kind of summed things up. Our innate drives are at conflict with what is good for us.

Chapter 7

Population and Technology as Driving Forces

The Neolithic Revolution was an amazing, unique point in human history. Starting about 315,000 years ago when *Homo sapiens* evolved from *Homo erectus*, our human population grew and dispersed out of Africa. By about 12,000 years ago, humans occupied the habitable continents of the globe (Waters 2016, 2) and our population continued to grow.

Population growth was not a dynamic hunter/gatherers sought; individual hunter/gatherer clans managed their population size to be small, to keep their demands on the environment within the bounds of what was provided naturally and to be able to maintain healthy cultures.

Numerous hunter/gatherer clans made up regional hunter/gatherer populations, and those regional populations grew, slowly, as clans expanded and split into multiple clans, possibly to the point of saturating habitable regions with clans (Redman 1999, 93). When this occurred, hunter/gatherer clans would have found themselves in a situation unique in the history of our species: forced to migrate less or not at all for lack of supporting territory and therefore at risk of depleting resources available to them (Binford 1983, 213). If they knew that even one of their neighboring clans had a propensity for violent raiding, they knew that a smaller population was more vulnerable to attack (LeBlanc 2003, 69-73; Carneiro 1970, 9-10).

Hunter/gatherers typically have options in the face of resource scarcity: they are deeply knowledgeable about the resources in their region and can choose to harvest a less desirable resource if the optimal is not available. They can also harvest optimal resources more intensively than they would otherwise. And being nomadic, they can move to another location better endowed with the resources they need. Additionally, their population could split into smaller groups and disperse, each putting less strain on the environment.

Just like every other organism, hunter/gatherers responded to variables in their environment changing. About 10,500 years ago, regional hunter/gatherer populations faced a new variable: a crowded neighborhood. This must have been a time to consider their options. In addition to those listed above, humans brought their latent abilities in their hands, brains and skill at communicating. In a scenario where relocating to find a supportive niche wasn't an option and where they were at risk of attack if they stayed as a small group without much technology, the option of agriculture, sedentism, population growth and technology would have been good options to pursue in spite of the downsides. Economist Ester Boserup and anthropologist Mark Nathan Cohen each seem to refer to this point: population growth and resource scarcity prompting a shift to an agricultural lifestyle (Cohen 1977, 11-14; Boserup 1981, ix-5).

From a distance of 10,500 years, we can't know precisely what hunter/gatherers were experiencing or how they weighed their options in the face of higher regional population size and increased density. But we do know that their response was unique: for the first time in the history of life on earth, an organism responded to changes in their environment by creating complex technology and seeking population growth. Our human ancestors made a choice that may or may not have been monumental at the moment, but we can look back and see the shift to agriculture as an incredible pivot point in the trajectory of humans. After hundreds of thousands of years of surviving within the biological system that limited their population size, at the Neolithic Revolution humans stepped into uncharted territory.

The Neolithic Revolution reflected our capacity as a species as we started to use the physiological traits we had evolved. We shifted focus to create and rely on technology, and now technology defines much of our lives. Our latent ability to create complex technologies emerged at the Neolithic and would dictate a path into our future. We have been on that path ever since. We are a uniquely technological species.

In assessing the shift humans made from being hunter/gatherers to growing their food, what was the driving, foundational factor that caused the shift? Many different factors influenced human clans and regional populations 10,500 years ago: their populations were growing slowly; as hunter/gatherers they created and used little technology; being human, they had the ability to be aggressive and knew others of their species could also be aggressive; as biological organisms, they needed particular resources to live; the resources they needed to live were of finite supply; and they brought uniquely evolved physiological traits to the business of survival, traits that allowed them to design and create complex technology.

Of these, population growth and the ability to create technology were the factors that drove the shift to agriculture. Without these two factors, which mutually influenced each other as regional populations approached the shift at the Neolithic, the human species would have remained like every other organism on earth – regional population sizes would have grown and contracted based on the vagaries of changes in the environment, at any given moment maintaining a population size based on the limiting factors that came to bear at that moment.

Instead, growing populations and evolving technology prompted each other to continue growing and evolving. This was a new path, and resulted in new consequences for the human species and the natural world.

Chapter 8

Environmental and Cultural Changes from Living Above the Natural Upper Limit

The change to agriculture involved a significantly different lifestyle from what hunter/gatherers lived. At the Neolithic people committed to agriculture, became sedentary, built permanent structures to live in, grouped together in settlements and eventually formed large, densely populated cities. With this shift, people also committed to a technological lifestyle – creating tools for agriculture, but also tools that made lives easier. Over time, we created technologies of roads and carts, looms and woven fabrics, we learned to smelt ores to make various metals, we created different materials such as paper, glass, rubber, plastics, and we learned to make a myriad of objects that made our lives easier. Technological innovation is embedded in human history since the Neolithic.

We developed and applied technology with particular goals in mind – to grow enough food to feed our population, to direct water to settlements so people and animals could drink, to create shelters to protect us, and many items that allow us to be more productive with less personal energy expenditure – and those technologies were effective in reaching those goals. But with these technologies and changes in lifestyle came secondary effects. Large, agricultural populations and the complex technologies that emerged with them generate additional dynamics and by-products. In the application of technology, these dynamics and by-products were not sought, but they come with the shift; they are inevitable.

Two areas of secondary effects from embracing technology and growing our population are important to us now: the effects humans have on the environment and the effects humans have on ourselves, our culture. Our relationship with the environment and our relationships with each other changed when we shifted to an agricultural lifestyle.

Environmental Effect of Technology and Population— Pollution

Two environmental problems emerge from our development of advanced technology and our large population: pollution and resource depletion. This section will explore the generation of pollution by humans.

To create and use every technology it is necessary to harvest and refine natural resources, manufacture the technology from those resources, apply the technology and eventually discard it. From any or all of the steps in this process, by-products result. A by-product is any secondary product not needed for the functioning of the technology or for the resolution of the problem the technology was created to resolve.

Hunter/gatherers created and used minimal technology made of minimally refined natural resources: rock for hunting, cutting, chopping and grinding tools; wood, plant fibers, bone, antler, skins, leather and fire for hunting, digging, sewing, clothing and cooking. These simple technologies were routinely applied in hunter/gatherer life, but in their fabrication, use and discarding, these technologies left no long-term by-products harmful to people or to other organisms.

As an example, a stone is a natural raw material that was shaped and altered to make a point for a spear or arrow. To make a point, a piece of stone is knapped, reduced in size and to an appropriate shape by using hammer stones, pieces of antler or bone and perhaps a piece of leather to support the point as it's being worked. The by-products from the process of knapping are stone flakes not needed for the desired point and the stone, bone, antler and leather used to fabricate that point. Archaeologists have identified many sites around the globe as quarries and manufacturing locations where hunter/gatherers accessed the type of stone they wanted and worked the raw materials into points. Hunter/gatherers from at least 3.3 m years ago have been shaping stone into points.

Technologies build on what came before, and our technologies and our skills in making them have evolved significantly from our creation of fractured rocks 3.3 million years ago. When an idea in the form of a product that will help people is envisioned, developed and comes to the market, it's critically evaluated – if people find it beneficial and cost-effective, it's accepted and is integrated into people's lives. As well, once available to the market, people explore it and work with it, to see if it can be applied in different ways or modified to create new technologies (Flynn 2007, 1-2).

Given the enormous variety of technologies created by humans, we produce a wide range of by-products. Some are innocuous; some are harmful to life processes. The questions are really what the by-products are, how much of them there are, and the impact they have when incorporated into biological systems. By-products that are toxic to life are pollution. Different pollutants vary enormously in their makeup and their impact on life processes.

Broadly, there are two ways pollutants can hurt people and biological systems. One is toxicity: when toxic by-products are absorbed, they disrupt normal, healthy functioning of our bodies and cause illness. This can happen in different ways: some toxins damage DNA, some attack organ functions, some disrupt hormonal systems in our bodies, some disrupt our nervous system. Some are highly toxic in small amounts; some persist for a long time in the environment; and once ingested, some remain in our bodies for long periods. There are many different ways that our bodies and pollutants interact and there is no one mechanism of deleterious effects. Examples of toxic by-products are tailing piles at ore mining sites that produce sulfuric acid; PCBs released to the environment; mercury released from coal-burning power plants; persistent organic pollutants spread into the environment; medicines that are flushed through sewage systems and released into the environment; or chemicals that are spread on fields that find their way to aquifers and drinking water supplies.

The second way pollution can hurt people is through sheer volume. When something is released to the earth's air, water or soil that changes biological functioning by an outsized presence, that is a form of pollution. Examples of large-volume pollutants are CO_2, water vapor, nitrogen oxides and unburned hydrocarbons emitted from internal combustion engines that change the planetary ratio of gases in the atmosphere; plastic pollution on land and in waters; SO_2 released from coal burning electricity-generation, generating acid rain; chlorofluorocarbons affecting our atmospheric ozone layer; fertilizer runoff into waterways causing dead zones in oceans; and light from cities. By their large volume, these by-products disrupt the functioning of natural biological systems.

The qualitative differences between by-products generated by hunter/gatherer technologies and by-products of recent human technologies are telling. By-products that hunter/gatherers left behind have for the most part decomposed through natural, biological processes: the wood, leather, bone, fiber tools they made were broken down by natural microbial decomposition. The more decay-resistant stone tools and the by-products of small stone shards have remained. Hunter/gatherer's technological by-products are neither toxic nor of large volume; they are not harmful to biological processes. However, the generation of more recent human technology, especially since the Industrial Revolution, produces by-products that are both toxic and of large volume.

In viewing the history of pollution, it is difficult to pinpoint when the qualities of by-products shift them from being benign to being polluting. Every technology, every by-product, all natural resources, every manufacturing process is unique, so it's impossible to assess the trends that have emerged over time and make an accurate statement about when and why our technologies became toxic to us. In general, it looks like the development of complex technologies is the point that their by-products proved to be problematic, and that the more we refine resources the more toxic the by-products become.

The Neolithic was the point that human technology began to create pollution that affected the environment and people's lives. Just as our population grew slowly over time and our technologies evolved gradually, our environmental impact trends showed up over long periods. Growing crops in some regions required irrigation which led to soil salinization on some fields (Markham 1994, 3). Domesticated animal grazing and overharvesting of trees in the Levant led to severe soil erosion in some areas (Redman 1999, 107-110). When humans stopped being nomadic and began living in settlements and later cities, concentrations of human and animal waste probably contaminated water supplies (Barnes 2005, 58; Markham 1994, 1-2). The nature of water-borne microbial diseases wasn't well understood and improper disposal of human wastes continued to cause disease in human populations until about the late 1800s (Markham 1994, 6-9). After humans learned to smelt metals from mineral ores, we created metal pollution: Icelandic ice cores dating from about 500 BCE reveal copper pollution (Hong 1996, 246-249) and a Serbian peat bog records lead pollution from about 3600 BCE (Longman 2018). We have found copper pollution, probably from an early form of copper smelting, in a riverbed in Jordan, dating to about 5000 BCE (Grattan 2016). Roman use of lead to store food resulted in lead poisoning (Kates 1985, 1). Humans clearing forests starting about 6000 BCE led to the increase in the level of atmospheric CO_2, and the increase in the level of methane starting about 3000 BCE is thought to be due to widespread cultivation of rice (Ruddiman 2003). Extensive burning of fossil fuels is blamed for more recent changes in atmospheric levels of CO_2.

The acceptance of a different lifestyle – agriculture – was an enormous change for humans. All the specific elements of this new lifestyle need to be viewed together, because it is together that they had an impact on human life. Shifting from being nomadic to being sedentary; from having minimal shelter to building permanent structures; from maintaining small populations to allowing growth; from maintaining harmonious cultures to allowing cultures to be chaotic and deleterious; from having simple economies to creating specialized and complex economies; from maintaining simple technologies to generating the evolution of complex technologies; each of these elements of our agricultural lifestyle influenced the others, and influenced how we related to our environment and related to each other. As well as the positive, desired benefits of technology emerging in this shift in lifestyle, we produced their by-products, some of which are harmful to biological life. Over time, the negative elements have become more evident and have resulted in the environmental degradation that we are staring at now.

Two observations are critical in this section about pollution. First, every human technology generates by-products, and some of those by-products are pollution. Second, every pollution we are concerned about now is a technological by-product. These two observations, viewed together, are deeply significant. Humans are the only organism in the world to create pollution to the degree that we do, pollution that is toxic, widespread and long lasting. With the creation and application of complex technologies, people have been and continue to make environmental problems for themselves.

Environmental Effect of Technology and Population— Resource Depletion

In addition to pollution, resource depletion is another secondary effect of a large population and advanced technology. All resources are scarce. With a population size above the natural upper limit we need to harvest more resources than the earth would produce and regenerate naturally. In addition, when we shift to using technology to support our population, resources need to be harvested to produce and maintain technologies upon which we are now reliant.

As hunter/gatherers, humans did not have a resource depleting impact on the natural world. Their lifestyle was one of maintaining small populations and migrating when necessary, to avoid overdrawing from their environment. This lifestyle is similar to that of all other organisms on earth, and is one that stays in synchrony with the biological systems of the world. Barring any catastrophic change to the environment – severe drought, volcano, hurricane – that precludes accessing resources needed to live, a population that does not over consume what it needs to live can survive for the long-term. This system has worked for millions of species for millions of years.

Human hunter/gatherers prior to the Neolithic Revolution probably had a resource-depleting impact on their environment in two ways. It appears that humans contributed to the extinction of megafauna around the globe, including mastodons, mammoths, ground sloths, horses and camels, using hunting technology of stone-tipped spears and arrows (Sandom 2014; Martin 1973, 969). Archaeological evidence is not conclusive, and it is possible that a changing climate was stressing megafauna at the same time that humans hunted them. But it is likely, even with relatively small regional populations, that human hunting contributed to depletion and extinction of some species.

The second way pre-Neolithic humans may have influenced their environment was through extensive burning of the landscape. Again, it's difficult to know if humans were responsible for the evidence of fire that shows up in archaeological sites, but interpretations currently weigh toward large-scale, intentional burning of landscapes by aboriginal peoples around the globe (Scott 2016, 5; Bowman 2011, 2224-2225; Gowlett 2016, 1-4). Humans are the only organism able to manage fire purposefully, and they used it long before the Neolithic.

Humans became capable of depleting resources when they shifted to agriculture at the Neolithic. In this shift, people reduced the consumption and growth constraints that hunter/gatherers lived by. Humans began to consume not only to maintain their lives, but also to support the new system of agriculture and to grow their population. People's relationship with the environment changed: our view shifted from seeing the natural world as a source of life-supporting resources to a place to harvest resources regardless of the consequences. Working with the environment became less important than taking what was wanted. It is likely that human's initial Neolithic depletion impact was slight: their population was relatively small, probably between 5-10 million people, and the scale and complexity of their technology was low, implying that their ability and need to overharvest resources was low. But the Neolithic involved a significant change in how people related to the resources of the natural world.

The Neolithic economy was still energized by daily solar input that grew plants as food: for direct human consumption, for consumption by the domestic animals working for people, and for consumption by the animals people still hunted for food. The applied technology of agriculture and related tools was a significant shift, but the population size did not grow dramatically at the Neolithic and did not have an outsized impact on resource consumption.

Charles Redman makes the point that after the Neolithic, humans could have a depleting effect on local animal populations in three ways: by actively hunting them for food or because they threatened domesticated livestock, by converting natural landscape to agricultural land and depriving indigenous animals of habitat they needed to survive, or by introducing invasive species that could either outcompete indigenous animals or could prey on them (Redman 1999, 56-62). Humans with hunter/gatherer-level technology also had the ability to deforest areas, to the extent of causing localized human population collapses (Railey 2008, 11-12; Redman 1999, 71-73).

Over time, the human population grew, technology evolved and the economy expanded, necessitating larger draws of resources. Until about 1750 CE, humans relied on muscle power from people and domesticated animals, wind and water power, and burning wood to produce heat to fuel their economy. As early as 1300 CE, wood shortages and deforestation were occurring in Europe. About 1750 CE Europeans turned to coal as a replacement fuel. Coal changed the game entirely: it provided a dense energy source, allowing more productivity in the economy. At the same time as coal usage began in earnest the European population increased, agriculture expanded and the steam engine evolved. The Industrial Revolution was a point in time that economic productivity and resource consumption expanded significantly.

With the expansion in population, the evolution of technology, the use of coal as an energy source and growth in the economy, human impact on the environment became significant. Just before the start of the Industrial Revolution, we depleted forests in some European regions and burned coal to toxic atmospheric levels in London. For an interesting read about the impact of early coal burning, look up *Fumifugium*, written in 1661 by an inhabitant of London (Evelyn 1661). It gives insight into the pollution and personal discomfort caused by burning coal.

The emergence of our industrial economy, energized by the use of fossil fuels, meant that we had crossed a tipping point of technological complexity. Humans were having a detrimental impact on the environment with our technology, its by-products, and a growing population. We were creating pollution and depleting resources.

Humans are currently consuming several resources faster than they can be regenerated, including water, soil in agricultural regions, some fish species, wetlands, some regional forests, and we're depleting biodiversity of the earth by causing rapid species extinction. As well, we are depleting atmospheric oxygen in some regions due to excessive pollution we generate; atmospheric pollution reduces oxygen availability to us. We are also depleting petroleum and mineral ores, which take millions of years to regenerate. In addition, our development of land depletes the earth of natural habitats (Foley 2011, 337-339; Lambin 2001; Agnoletti 2014, 79-85; Allsopp 2009, 33-43; IPBES 2019, 10-16).

Cultural Effect of Technology and Population— Disharmony

Hunter/gatherer clans were nomadic, they kept their populations small, had no central leadership, all individuals were treated as equal and they resolved conflicts between individuals publicly, with the whole clan weighing in. Reciprocity and sharing were part of their economies; even in larger and more complex populations with head-men or leaders, ceremonies of giving to the leaders were expected to be followed at some point by giving the gifts away. When someone within the clan came into a significant food source, it was shared equally among the whole group (Hitchcock 2000, 19; Lee 1999, 4). These behaviors typified regional hunter/gatherer populations around the globe. Independently, different populations produced approximately similar patterns within their clans.

These behaviors were enormously significant: they had the effect of helping to reduce divisive elements within the clan by minimizing the emotional trauma of prolonged personal conflicts and by minimizing the impact of inflated egos. By living in small groups, by discussing conflicts openly, by each clan member knowing each other closely, by the lack of hierarchy and status, individuals were self-constrained when they felt like expressing anger or harmful emotions. If a conflict did come up, the whole clan was involved in its resolve and in maintaining peace. And without central leadership or social hierarchy, with sharing and reciprocal giving, with few possessions, hunter/gatherer cultures were egalitarian. No person was above or below another (Lee 1999, 3-4; Tainter 1988, 35-36). Individuals were not singled out as being better than others because that would create status and hierarchy.

By minimizing the negative impacts of interpersonal conflict within the clan and by maintaining equality among the clan's individuals, clan culture nurtured stability and longevity. Given human nature and the potential for anger, conflict, and inflated egos to cause divisions within the clan, clan culture was a powerful way to favor survival for the long-term. Checked egos and resolved conflicts benefitted all individuals. The clan thrived when all worked together for the common goal of survival (Lee 1999, 4; Sahlins 1972, 1-39). Band culture is a form of 'survival of the fittest' – humans are social creatures, and individuals are likelier to survive when they work together.

Hunter/gatherers were aware of not overdrawing resources from the land they lived in; by managing consumption, they generated a long-term relationship of supply and demand with their environment. In a parallel way, by the behaviors they brought to their culture they generated close and long-term relationships with each other. Whether by intuition or conscious design, hunter/gatherers knew to structure their cultures to avoid the dysfunction that comes with unresolved conflict and inequality. In the relationships they maintained with the environment and with each other hunter/gatherers lived sustainable lifestyles, sustainable in the ecological sense of the word. Their lifestyles could be maintained like this indefinitely.

The shift to agriculture changed the form of human culture and changed how people relate to each other. With agriculture populations became sedentary, local settlements formed, complex economies developed, and populations grew. More importantly, not only did agriculturalism involve these different lifestyle traits, it led onto a path of growth: population growth, technology evolution, economic growth. Shifting from the culturally stable lifestyle of hunter/gatherer to the agricultural lifestyle brought deep changes to human culture – how people relate to each other.

Anthropologist Robin Dunbar has researched to discern population sizes that are the most stable and productive for humans. He found that groups of approximately 50, of 150 and of 500 tend to stay together longer than other sizes, based on his observations of contemporary groups of people. Dunbar does not explain precisely what factors in our psyches prompt our tendency toward these sized groups, but believes it is likely that hunter/gatherer clans were structured in response to the same group dynamics (Dunbar 2017). What Dunbar's research implies is that 50 people is one optimal size for hunter/gatherer clans, and that growing a population beyond 500 would result in less harmonious and stable populations.

The cultural harmony that is possible in a 50-person clan is difficult to maintain in a much larger population. The change at the Neolithic to larger and denser populations and the hierarchy that developed from political and economic complexity meant that individuals could no longer know each of the others in their settlement on a personal basis or maintain close relationships with everyone around them. These dynamics made it more difficult to create group cultures to be cohesive, personal and harmonious (Wirth 1938, 11-17). In large populations, individuals are more anonymous and have impersonal connections with many of the people around them.

It seems to be that looking out primarily for one's self is a default position wired into human nature. Even without prompts, people tend to follow a path to accrue power, gain wealth, or enhance the ego (Hayden 1995, 20-21) although an internal compass may guide an individual away from a self-serving direction. More often, cultural constraints and principles come to bear. We cannot preclude the less harmonious aspects of human nature from our beings, but culture can help mollify the negative emotions and their expression. In a small population, close and personal relationships help generate equality and harmony between individuals and those close relationships generate group cohesion and support. In a large population close and personal relationships are less frequent; anonymity and distance between individuals are more the standard. A large population can design and structure its rules to attempt an equal and harmonious population, but when cultural norms that inhibit self-centeredness are avoidable or ineffective, self-regard and inequality seep in.

It seems that in large populations, two categories of divisive behavior emerge, each resulting in disharmony within the culture.

Disharmonious Behavior

In a large population, the cultural harmony generated by the close relationships of small populations can easily be lost. To establish desirable cultural norms for the whole group, some sanctioned entity has to design, apply and enforce those rules. After the formation of governments in about 5,000 BCE, the state was empowered to apply cultural rules. But in designing, applying and enforcing a culture's rules and in seeking to shape individual behavior and maintain harmony within the population, the state is faced with a different dynamic than a small population. The same issues of human nature bubble up, but because of the size and complexity of the population and the distance between the state and each individual, management is more difficult. In a small population, individuals are constrained from harmful behavior by knowing they have to interface closely, daily, with those in their clan, and if they do hurt someone close, they will face personal repercussions from people in their immediate circle. These are people relied upon for food and emotional support, people who help them survive.

In a hunter/gatherer population, the clan is the creator and enforcer of rules. In a large population, the state is the creator and enforcer of rules. But the relationship of the state to individuals of the population is much more distant than the relationship of the clan to its individuals. Simply knowing that there are not close, personal connections watching all interactions may make someone prone to act out less positive aspects of their nature.

Large populations face inherent difficulties in maintaining harmony and equanimity over the long-term. The very nature of being large means that populations automatically create dynamics of distance between individuals, which leads to different expressions of our human nature than in small, close populations. Those different expressions are more frequently of the negative sort – for example expressions of anger, dissention, theft or criticism. Individuals think twice about acting in these ways in a small, close group. In a larger population with more distant relationships these behaviors are less frequently squelched. This dynamic is one of the significant effects that were the result of shifting to agriculture, large populations, complexity and growth. Population cohesion and equanimity over the long-term are much more difficult.

Cultural Inequality

Besides allowing the expression of disharmonious aspects of our human nature, large populations generate inequality. Anthropologists recognize the presence of inequality as a significant dynamic within a population. One thread of research believes that inequality emerged with the transition to agriculture and complex societies: hunter/gatherers were egalitarian, but agricultural cultures developed inequality (Price 1995, 129-130; Inglehart 2016, 2-3; Alacevich 2018, 13-14). Observing similar dynamics, economist Angus Deaton writes that inequality between countries and between people within countries emerges as a result of human progress, and that inequality increases when progress occurs, such as during the Industrial Revolution (Deaton 2013, 1-6). Gary Feinman questions the premise that inequality emerged starting at the Neolithic, pointing to evidence of inequality in hunter/gatherer bands. Feinman does not believe agriculture is a prerequisite for inequality (Feinman 1995, 255-259).

While small hunter/gatherer populations are functional and healthy without central leadership or political structure, larger populations generate some form of leadership to manage the complexities that come with size, and with that political structure comes political power and social hierarchy. As well, large populations prompt the generation of larger and complex economies, leading to specialization of production roles and therefore hierarchy within the population. With hierarchy comes inequality.

With a group as small as two people, there are bound to be inequalities between them. One will run faster than the other, one has better eyesight, one is a better hunter, one has the nature of a leader and another wants to be supportive. So even in the small group size that hunter/gatherers maintain, it is inevitable that inequalities exist between members.

Feinman's observation that some hunter/gatherer populations had aspects of inequality is consistent with the inevitability of differences between people. All populations, small or large, inevitably contain inequality because people are different, but the degree of expression in a culture is variable and manageable. In hunter/gatherer clans, the whole group set the rules defining the expression of detrimental emotions and resource allocation, and maintained a collective awareness that cohesion meant survival (Hayden 1995, 20).

Expressed inequality in a population indicates that some group has access to fewer resources or less power than others, and because of this can engage less fully in the activities of the group. In this way, inequality is a divisive element – those with access to fewer resources feel left out, deprived, less accepted. Inequality has a powerful and negative impact on how individuals within the population relate to each other (Crow 2011, 4-5). Hunter/gatherers knew the divisiveness of inequality is harmful for the long-term health of their group, so they designed policy to minimize it.

In the bigger picture, entrenched inequalities imply a deeper disharmony within a culture. Established inequality indicates that the rule-setting body of the population is out of touch with what it takes to maintain equity and therefore harmony and cohesion within the population. If the political structure of a culture sanctions inequality, the culture excludes some people. When the bigger picture of group harmony, cohesion and long-term survival undergirds policy, policy-makers design all policy to those ends, and equality will emanate from specific policy designs. In their book *The Spirit Level*, Richard Wilkinson and Kate Pickett point out that the best way to reduce the harm caused by cultural inequalities is to reduce inequality. Putting band aids on the problems that emerge, such as therapy for feeling excluded, does not address the core problem (Wilkinson 2010, 33). Policy-makers may defend policies that maintain inequalities for various reasons, but what that points to is that policy-makers seek the outcomes their policies generate more than they seek equal access to resources for all in the population and long-term stability.

That our *Homo sapiens* hunter/gatherer ancestors survived for 300,000 years is no small matter – humans are complex creatures with complex and variable human natures. Sometimes we behave in ways that draw people together; sometimes we behave in ways that actively create disharmony and division. The cultural structure hunter/gatherers designed allowed them to survive for a long time – for hundreds of thousands of years. Harmony and equality must have been driving the rules of their culture. This is immensely informative. By strict sharing of food, by maintaining a small clan size that didn't need central leadership, in not praising successful hunters, in not accumulating goods, by resolving conflicts publicly, our hunter/gatherer ancestors knew to live by rules that reduced or eliminated divisive elements between clan members. At a functional level, they understood that negative behaviors and inequalities were divisive and detrimental to long-term stability and survival (Tainter 1988, 36).

The existence of inequalities within current cultures is extensive; inequality takes many forms. There have been income inequalities at the global level for the past 200 years, both between countries and within countries (Milanovic 2011; Alvaredo 2017, 8-20). Within countries, individuals have unequal access to education and health care (Neckerman 2007). Income inequalities often underlie the generation of other inequalities, since income allows access to beneficial elements of life (Crow 2011, 9-95; Wilkinson 2010). Racism and sexism are inequalities that impose hardships on people in many countries.

Economist Thomas Piketty and colleagues have done insightful research about income and wealth inequality within our relatively recent population. They have compiled and analyzed data from various countries; some of their data sources go back as far as 1700. Coming from a social science and economic perspective, Piketty observes that inequality is an inherent element of capitalist economies (Piketty

2014, 20-27) and that the way to counter inequality is with purposeful policy that stimulates a more equal distribution of resources. A policy of prompting an economy toward equality has to be a deliberate choice by a culture's policy-makers.

The important point here is that large populations are inherently complex and inevitably generate inequality, and without a clear, direct, intentional goal on the part of policymakers of the state toward reducing inequality within a population, inequality is likely to emerge (Rosanvallon 2016) inevitably prompting divisiveness with the population.

Chapter 9

Potential of Collapse

At the Neolithic, our population shifted to an agricultural lifestyle and our population grew above the natural upper limit. Consequently, the large population is vulnerable to a shock if that supporting agriculture becomes unavailable for any reason: pests, drought, flood, unseasonal freezing, decimation from a raiding neighbor or mismanagement of the technology. Loss of agriculture or any supporting technology brings the risk of societal collapse.

Societal collapses have happened many times in the past: the Maya civilization of Central America, the ancient Roman empire of Italy, the Harappan civilization in what is now Pakistan, the Moche culture in Peru, the Old Kingdom of Egypt, the Akkadian empire in Mesopotamia, the Minoan civilization in the Mediterranean, the Norte Chico civilization in Peru, the Ancestral Puebloan culture in southwestern North America, Cahokia in the Midwestern United States, the Norse settlements on Greenland, the Khmer Empire in Cambodia, and many more. These cultures all formed after the Neolithic, evolved to be complex, and collapsed.

There is strong academic interest in understanding collapses, and there are different definitions of what a collapse is. One way to define collapse is a significant population decline (Diamond 2005, 3). Another view is that collapse involves a loss of cultural and political complexity (Tainter 1988, 4). Glen Schwartz considers collapse to be "… the fragmentation of states into smaller political entities; the partial abandonment or complete desertion of urban centers, along with the loss or depletion of their centralizing functions; the breakdown of regional economic systems; and the failure of civilizational ideologies" (Schwartz 2006, 5-6).

When viewed together, these perspectives on collapse come from the same point. Whether analyzed through population size, political structure or an economic lens, a collapse involves some form of retrenchment of a population from the level of complexity it had evolved to over time. Populations gain complexity after becoming large and agricultural (Carneiro 1986) and some degree of that complexity is reversed by a collapse. The culture and the population may not disappear entirely, but at the least it shifts to a fewer number of parts and is less complex.

Archaeologists and anthropologists have researched and written about causes for specific collapses and have questioned why collapses happen at all. From a distance, it is difficult to know what factors pushed a culture into collapse; clear consensus is lacking. Relevant factors were numerous: political upheaval or leadership dysfunction, local and regional weather changes over both the short- and the long-term, nutrient loss and salt build-up in agricultural soils, water availability for both crops and direct human consumption, and raiding by outside forces. And it is likely that several simultaneous factors pushed a culture to collapse. The dynamic of resource depletion comes up persistently in explanations for many collapsed peoples. In Joseph Tainter's review of different authors' analyses of collapse involving resource depletion, he noted that either human mismanagement or macro changes of the environment depleted resources (Tainter 1988, 43-51). Archaeologist Steven LeBlanc makes a strong case that resource depletion prompts aggression and warfare by one population on another (LeBlanc 2003, 69-71). Warfare is not the same thing as population collapse, but each is likely a function of a population impinged by significant stress.

It is difficult to know if lack of resources caused partial or complete collapses in the past but we can be sure that the need to have adequate resources available to them was just as acute as it is for us now. At the least, we need food, water and air. When our population grows above the natural upper limit the risk of finding ourselves without those resources increases significantly: our rate of consumption is above what the earth can regenerate sustainably, so we become dependent on the technology that goes beyond the natural biologic systems of the earth to harvest what we need. Whether or not we see the risk in the midst of our relative resource abundance, it is there.

The dynamic of living above the natural upper limit – creating and relying on technology, managing the supply of resources needed to survive, generating complex economies and complex cultures – is unique to agriculturalists. Hunter/gatherers did not face the risk of resource depletion on a regular basis because they can be flexible in how and what resources they harvest. The natural world supplied what they needed to live, and they knew how to harvest what they needed. Agriculturalists do not have that flexibility. The aspect and risk of resource depletion is unique to our complex societies.

Chapter 10

Technological Complexity

From the time the first technology was created and used, as long ago as 3.3 million years ago when stones were fractured to create sharp-edged tools (Harmand 2015) up to the Neolithic, all hominin technology was at essentially the same level of complexity. Knapping stone points and stone tools, making clothing from animal skins, crafting spears and bows from wood, shaping bone and ivory for fishhooks and needles, and using wood and antlers for digging tools meant using readily available natural resources and modifying them minimally with simple tools, using human muscle power. The creation and use of these technologies and the generation of by-products did not harm biological systems.

Since the Neolithic, our technologies have evolved, typically building on technologies that came before. Each technology addressed a specific problem. Different materials were refined for application in a technology; some technologies are human-powered and some are externally powered. How long it functions and how long it takes to break down after being discarded are unique to each technology. It is unrealistic to make direct comparisons of one technology to another, but we can generalize by observing that as our technologies have evolved, they have become more complex. And with complexity they and their by-products have gotten more toxic (Harriss 1978, 6-9).

Complexity involves more steps, more refining of natural resources, more parts and more interactions between parts in the process of designing, creating and using technology. Where a hunter/gatherer might heft a rock as a tool to pound something, modern humans make a hammer out of wood and steel. The processes of mining and refining iron ore, forging a hammerhead, harvesting the tree, shaping the handle and putting the two together are more complex than picking up a rock, and the by-products of those processes are more deleterious than a discarded rock hammer. The hammer is a more complex technology.

It is hard to say specifically why our evolved and more complex technologies generate more pollution. To create our first technologies, initially all we had were the natural resources around us to work with, and we used only an intuitive understanding of those materials and minimal skills. Over time we explored the properties of those materials, experimented and observed their composition in finer detail, and so came to deeper and more nuanced ability to use them. As well, we gained more experience and skills in making tools to form and manipulate resources. When new problems bubbled up, we applied more refined resources, modified and formed with nuanced skills.

The materials for all technologies originate from the earth's natural resources. But for some reason, complex refining and modifying resources exposes the toxic qualities in them. In the evolution of our technology we seem to generate toxic by-products the more we refine, combine and manipulate resources. This is not a hard and fast rule, but a broad observation. In the creation of every technology some form of by-product is inevitable; with highly refined technology those by-products tend to be complex, toxic and deleterious to human health.

Every resource is different as is the way we manipulate each resource, so over the history of technology there was not a point in time where the by-products went from benign to toxic. Prior to the Neolithic the simple technologies of hunter/gatherers were not polluting and were not evolving – hunter/gatherers could have kept on their track indefinitely without creating pollution. At the Neolithic, this changed.

Our technologies were not complex at first but they evolved to be complex. The application of our technologies has both positive and negative effects: they resolve problems they were created for, but they also introduce hazards, both environmental and cultural, that negatively affect humans (Kates 1985; Harriss 1978).

Chapter 11

Energy Sources

Before 10,500 BCE, humans hunted and gathered food and water, migrated, built shelters and crafted tools and clothing. These actions helped them survive; and their bodies, their muscles, did the work (Sahlins 1972, 5-6; Zalasiewicz 2008, 5). Solar radiation fueled our ancestor's actions: they ate plants and animals that lived within the biological system of life fueled by energy from the sun. Food gave their muscles the energy needed to do work. All organisms of the earth, including humans, gain energy from sources that link back directly to solar radiation (Smil 2017, 225) except for a relatively small number of organisms in the deep ocean who synthesize energy from chemicals (Keeton 1972, 636). The sun in our solar system provides the energy that drives all life on earth.

Human's first technology harnessing energy external to them was fire. Wood and plant material are commonly available biofuels, they burn easily to release stored solar energy, they are simple to ignite and can be harvested readily. It is difficult to know when humans learned to start and control fire for their benefit; archaeological evidence can be erased easily by natural forces, and it is difficult to know whether a burn layer in an archaeological site is from human fire or from a naturally occurring fire. However, hominins have probably been aware of fire for 2 million years, starting on the savannahs of Africa. Fire is associated with human archaeological evidence starting 1.5 million years ago (Gowlett 2016) and there is archaeological evidence that hominins have been using intentionally set wood fires for at least 400,000 years (Roebroeks 2011). It is possible that hominin hunter/gatherers used fire for large-scale clearing of landscapes, although there is much debate in the archaeological community about the degree of intentional control of fire our ancestors maintained (Bowman 2011, 2223-2225).

Primatologist Richard Wrangham postulates a significant aspect to the human use of fire: that hominin's ability to harness fire, starting with *Homo erectus*, gave them nutritional benefits that came with cooked food, resulting in evolved changes to human physiology of shorter guts and larger brains. The control and use of fire may have also resulted in decreased predation of humans by predators and decreased incidence of disease (Wrangham 2010). The technology of controlled fire use led to very personal biological advantages in humans.

Most hunter/gatherers processed their food minimally, and then consumed it with little time lag. Starting at the Neolithic, our populations grew and our technology evolved, and our methods for harvesting foods that supplied our energy needs changed. Agriculturalists grew more than needed at any given moment and learned to store what they grew to draw on throughout the year until the next harvest. They were effectively storing solar energy in the form of grains.

At the Neolithic, humans changed lifestyles in several ways – one change was the willingness to create and apply technologies to help them be more productive than their hunter/gatherers ancestors. Among other technologies, agriculturalists learned to gain energy from sources external to them, not to eat, but to do work. The following information about the history of energy use starting at the Neolithic is drawn from the *Handbook of Energy* (Cleveland 2014).

After the Neolithic, wood was a primary source of fuel for heat, to the point that in some regions trees were overharvested. By 2000 BCE, the Indus Valley civilization suffered deforestation, possibly contributing to its collapse. By 380 BCE Greece was deforested in areas, and

Rome by the mid-100s BCE; both areas used wood for heat, smelting of metals and shipbuilding. From the early 1300s to the 1600s, England and Europe experienced wood shortages, prompting an increase in coal mining in England.

Besides burning wood and biomass for heat energy, early civilizations created technologies to harvest water and wind energy. In India, the first known water wheel moved water to an irrigation canal between 350 and 200 BCE. By the early 1000s CE waterwheels of different designs were widely used in Europe. Into the 1700s, water was a significant source of power, energizing grain mills, water pumps and mechanical operations.

In 3500 BCE on the Nile River and the Mediterranean, we harnessed the wind to move boats. Sailboat technology for productive purposes evolved and improved well into the 1800s BCE. By 600 CE windmills were in use in Persia, China and Afghanistan, and by 1180 CE windmills were widely used in Europe. In China, windmills were in use in 1219 CE to grind grain and pump water. Technology continued to improve to make wind power more efficient, and into the late 1800s, wind power pumped water, milled grain, and generated electricity. Currently, wind is one of the renewable energy sources tapped to produce electricity on a large scale.

Human and domestic animal muscle, biofuels, water and wind all provide energy that is closely derived from solar radiation. Plants that are the basis of our food chain rely on solar energy to grow, as do trees whose wood we use for heat energy. Wind and water all move in large-scale patterns that are energized by solar radiation. In the same way that the sun is the foundation to all life on earth, solar energy has energized human productivity that allows humans to live and to grow our population. Before the Neolithic Revolution, we lived much like every other organism, living and responding to the patterns and dynamics of the natural resources available to us. After the Neolithic, we developed and applied our technologies to harness more and different types of energy – biofuels, wind and moving water – that allowed us to be more productive, but these sources were still directly dependent on daily solar input.

Our reliance on daily-solar-based energy sources changed as we shifted to using coal, a fuel naturally formed from huge volumes of plants growing millions of years ago being compressed within the earth. Coal has been used for heat in China and Wales heated with coal since 3000 BCE, and was in use in England and other areas in Europe in 1200 CE for metalworking. Technologies to both mine coal and to learn about coal's attributes progressed gradually until the early 1600s CE in England, when wood's scarcity due to overharvesting led to a need for a different fuel supply, and the English turned to coal. Wood shortages in various regions of Europe made the need more acute. The ability to mine larger volumes of coal helped prompt the Industrial Revolution, starting in about the mid-1700s.

Coal is widely used today as a fuel source to generate heat in electricity generation plants. It is mined around the globe, and is a commodity on the energy market.

Petroleum oil is another fossil fuel energy source humans utilize. Similar to coal, oil is formed from large quantities of biologic organisms, mostly zooplankton and algae, buried underground for millions of years. In 3000 BCE in the Middle East, seepages from petroleum reservoirs were used in roadbuilding, ship making, for medicines and other technologies, and by 2200 BCE petroleum oil was refined in China for heating and lighting. By the early 1700s CE oil products were being burned for heat and light. Throughout the 1800s, oil wells were established globally, and an oil market took form. In the late 1800s, development of internal combustion technology began producing reliable engines, able to transform petroleum oil energy into machines able to do mechanical work. Currently, global companies produce multiple kinds of fuels, plastics, chemicals and fertilizers for agriculture from petroleum oil.

Natural gas is a by-product of anaerobic decomposition of the plants and animals that form coal and oil, so natural gas deposits are often associated with coal and petroleum locations. Natural gas was used in 900 BCE in China and 191 BCE in Rome. In the United States, the first natural gas well was bored in New York, and used the gas to provide lighting and heating. (Cleveland 2014, 15-142)

This brief history of energy harvesting and consumption by humans is interesting but the bigger picture this history tells is vital. Hunter/gatherers used muscles to do all the work needed to survive; food supplied through the biological system of plant and animal growth based on daily solar input fueled the muscles that did that work. Hunter/gatherers also harnessed the concentrated solar energy of wood to provide heat, an energy technology they used to an evolutionary advantage by cooking their food. At the Neolithic Revolution, people still relied on muscles for physical labor and learned to domesticate animals for their additional muscle labor. For Neolithic populations, wood was still the primary external energy source, and besides for cooking, wood provided heat for firing pottery kilns. As technology developed over time, people learned to smelt ores to create metals, using wood to provide heat, and we learned to harness wind and water movement to gain mechanical productivity. Later still, at a significant transition, as we were overharvesting wood for metal production and home use, we turned to the concentrated energy of coal. Combined with other cultural elements that were going on at that time, using an energy-dense fossil fuel helped launch the Industrial Revolution. Coal was a concentrated energy we harnessed to replace muscles in our production systems and that energized the significant increase in productivity that came with industrialization. This marked a shift away from being part of the biologic system of utilizing daily solar input to energize work to a system using an energy dense fossil fuel (Smil 2017, 295; Livi-Baci 2017, 22-23).

The significance of this shift cannot be overstated. Muscles, whether human or domesticated animal, required fuel from food grown with the daily solar input that energizes the earth's biological systems. For hundreds of thousands of years, our species was one of many in this system. Our shift to using a fossil fuel to energize our economy took us out of that system and dramatically changed how we did our work. E. A. Wrigley states, "(C)oal offers a means of escape from the constraints of organic economies which photosynthesis does not" (Wrigley 2010, 22). The industrial system, growing and evolving from its start in about 1830, is founded on utilizing energy-dense fuel to power internal combustion engines and create electricity that energizes motors and drives the related technologies. It is an incredibly productive system. The energy-dense fossil fuels we use have both substituted for muscle energy to do work and added to the ability of our economy to be productive (Smil 2017, 5-8). With this use of fossil fuels, our population has grown, we have created more and more complex technology, and our economy has expanded.

For example, we travel and move large volumes of products using cars, trains, trucks, buses, boats, and airplanes powered by internal combustion engines. Our own legs, or those of domestic animals, did the carrying and pulling on farms and in factories before the development of internal combustion engines and electric motors.

Harnessing external energy to power technologies that increase our productivity not only does work that our muscles would otherwise do; we have developed technologies that do the work our brains would do. Digital technologies can process high volumes of complex intellectual information quickly and is powered by electricity generated by energy sources external to humans.

Fossil fuels were formed from the biological life system based on daily solar input but the concentrated nature of these fuels makes their consumption much different from the consumption of plants and animals on a daily basis. The rapid burning of fossil fuels not only releases the concentrated energy but also the concentrated components of those old plants and animals, including carbon, sulphur and mercury. And by burning fossil fuels in large quantities we generate combustion by-products in large quantities: nitrogen oxides, carbon dioxide, sulphur oxides and water vapor.

The use of fossil fuels to augment and supplant muscle energy is a very simple dynamic with enormous repercussions. The natural biologic system of life, consuming resources in a system based on daily solar input, is sustainable in the ecological sense. Life could go on in this system indefinitely: organisms living, consuming resources they need to survive, reproducing, dying, all at rates that neither over-consume nor pollute to the point of creating long-term or widespread damage. In creating technology energized by fossil fuels to do work for us we shifted from harvesting energy using the biologic system to harvesting energy using a system based on consuming concentrated but finite fossil fuels (Smil 2017, 295).

Because of human's unique evolution we have the ability to create technology that significantly changes our consumption patterns and consequently allows us to grow our population and consume resources beyond what the earth's biologic system can process. Along with the beneficial aspects of this consumption come the detrimental aspects of pollution from burning fossil fuels.

Chapter 12

Hunter/Gatherer Harmony v. Agriculturalist Violence

Anthropologists believe that the internal culture of hunter/gatherer clans was harmonious and cohesive, based on mutual support and equality among individuals. The picture that is drawn of hunter/gatherer clans is of a culture of working together for mutual benefit – each clan member benefits when they work together.

The behavior of pre-Neolithic hunter/gatherer clans to other clans is less clear. Lacking specific evidence it is difficult to know how clans in a region interacted with each other. Popular and academic belief used to be that hunter/gatherers were largely peaceful and non-aggressive to neighboring tribes, similar to bonobo chimpanzees. That understanding has changed. There is much archaeological evidence that clans fought with each other and that warfare, raiding and killing were frequent (LeBlanc 2003, 1-22; Kelly 2005). There are many archaeological sites of individuals and groups of people whose skeletons show evidence consistent with human-inflicted violence: skeletons with broken bones, projectile points embedded in the skeleton, groups of bodies not buried but left as they fell. We cannot know the specific dynamics of each site from thousands of years' distance but significant evidence indicates that warfare and aggression between clans was a consistent element of human life.

Besides LeBlanc's research, Steven Pinker's book *The Better Angels of Our Nature: Why Violence Has Declined* gives us insights into the human nature behind warring and violence by humans on humans. The dynamics are grim but important to recognize and bear on the topic of this book. Pinker makes a compelling case that our species is predisposed and psychologically motivated to be aggressive; that lesser angels deep in our psyches are a function of the evolutionary benefits of survival that come from defending resources we have against people who are similar to us (Pinker 2011, 4-40). The biological drive to survive and to see our genes carried forward is strong evolutionary impetus to aggress against others of our species with the same goal.

Intra-group behavior differs from inter-group behavior. How people see the individuals of a population they are close to and rely on for survival is different from how they see people in distant relationships. Within a clan, working together to harvest resources and share harvests benefits each clan member. In a similar way being empathic, helpful, supportive and caring of the group of people you are part of is evolutionarily logical. But people outside the clan are potential competitors: they have the same needs for resources and they have the same strengths and abilities to take those resources. Aggression seems to be an evolved, primal defensive position. On this basis, hunter/gatherers could have maintained harmonious cultures within their clan while being antagonistic and hostile to outsiders.

However, at the Neolithic Revolution the structure of the clan dynamic changed. As regional populations grew, people applied technology, economies grew and specialized, social hierarchy developed, and property ownership emerged. The economic and demographic changes would have led to individuals not having close relationships with all of those in their settlement and being resentful of others having more wealth or power. With the lack of close, personal connections the benefits of remaining civil became less evident and the self-serving aspects of our human nature less constrained.

It is believed that the first cities of the world formed about 4000 BCE in Sumer, in Mesopotamia in western Asia (McNeill 2003, 4). Other early cities formed later, in the Indus Valley of India and in China. The first cities were also the locations of the first forms of governments, generated to manage the activities and direction of the large and relatively dense populations.

The Better Angels of Our Nature makes the point that from the Neolithic Revolution until the formation of governments, regional populations of people were living in a time of minimal civility toward each other. Local populations were subject to consistent threat of invasion from marauding tribes or torture and death from militia beholden to local, powerful landholders or to sanctions from the church.

The formation of governments changed that. It was not to the government's advantage to have people preoccupied by the possibility of being tortured or killed simply out of malevolence. It was in fact in the state's interest to have its population being productive. *The Better Angels of Our Nature* traces the trends of violence within the human population and points out that with the formation of governments and the development of large economies, violence decreased. Governments created and imposed laws to define proper behavior, behavior that generated healthy economic activity. Governments benefitted and the population benefitted.

There is an interesting dynamic here: the state acted as a constraining element on the violent and malevolent tendencies of human nature, which is precisely what hunter/gatherer culture did. Hunter/gatherers maintained small populations, with internal dynamics unique to small populations; these small populations evolved the rules that worked for them to maintain harmony and to survive. The growth in regional population size meant that clan dynamics were not effective but the formation of states led to laws that had the same effect. Obviously, the ways that rules or laws are enforced differed between the hunter/gatherer system and the state-controlled system but the desired result was similar.

Chapter 13

Economics of Being Above the Natural Upper Limit

To understand the environmental and cultural impact humans are having, we have primarily been looking through the lenses of biology and anthropology to explain some trends and dynamics. There are interesting observations that can help us better understand the changed relationships we have with the environment and with ourselves. The study of economics provides insight into those observations. This is fascinating stuff.

Economics and Biology

Like those from biology and anthropology, economic concepts play a significant role in this book. Seeing how a region's economy evolves and how it becomes more complex give good insight into what we as a species are doing as our population grows. What is it about economics that bears on the dynamics of our population growth and our technology evolution?

"Economics is the study of how societies use scarce resources to produce valuable commodities and distribute them among different people" (Samuelson 2001, 4). The field has been evolving since Adam Smith, Robert Malthus and David Ricardo and others made observation on the market dynamics they saw around them in the mid- and late-1800s. Economics is a wonderfully deep, rich and complex field. It has a lot to say about the resources people consume to survive.

Population biology of hunter/gatherers is similar to that of any other organism on earth. Populations grow and contract based on environmental conditions including resource supply, predation and disease. And the economy of hunter/gatherers cannot be much simpler: a free market of the environment supplying natural resources and humans demanding them. Market dynamics change when a population shifts to agriculture. Past the natural upper limit market dynamics are more complex: they are affected by technology, extensive trade, continued population growth, the need to generate profits and multiple other factors.

Biology can be defined as the study of life (Keeton 1972, 1). Refining that a notch, we can observe that since the primary job of every living organism is to simply survive and to do so requires harvesting resources, then another definition of biology is: the study of resource transfer systems within the natural world. In the biological world, it is either eat or be eaten. We all need to take in resources, and we are all a resource to something else.

If we look at the definitions of both biology and economics, it looks like the two fields are similar: they both involve organisms acquiring resources for the purpose of survival. In fact, suppose biology and economics are the same: suppose economics studies the biology of humans – not all aspects, but the biological dynamics of resource harvesting and consumption for the benefit of human survival. Suppose the study of economics was only formed or generated by the growth in population size that led to an agricultural lifestyle and its inherent complexities. There is no need for economic study in a hunter/gatherer culture – the environment supplies or it does not; humans demand or they do not – that is too simple a system to require deep analysis. We study the microeconomics of organisms other than humans, but we call that biology. The study of economics is generated by our economy having elements of complexity: technology, extensive trade, property ownership, profits, pin factories, continued growth and many others, all economic elements that originated after the Neolithic Transition.

Economics Below and Above the Natural Upper Limit

Like every other organism on the planet, hunter/gatherers maintained a population size that could harvest the resources they needed to survive without depleting those resources to the point of extirpation. By harvesting some but not all, a resource could regenerate and be available in the future. This is an ecologically sustainable supply-and-demand relationship, one that could go on indefinitely.

Hunter/gatherers maintained relatively simply economies. Theirs was a pure free market economy. When the population demanded a resource, they looked to the environment to supply it. The relationship between environment and humans was simple, direct and clear, based on the microeconomics of supply and demand.

At the Neolithic, economic dynamics changed significantly. Agriculturalists actively manage their land to reap a benefit from it in the form of crops. Land management naturally leads to land ownership: after expending personal energy to gain a future return, it is rational to want some guarantee of having a right to that return. Land ownership and vested interests lead to concepts of profit and wealth; the realization that manipulated land produces more than untended land leads to the realization that more intensive manipulation will lead to more than is needed at the moment, which is profit that can be stored as wealth.

In this way, shifting to agriculture generates economic concepts unknown to hunter/gatherers: profit, wealth, property ownership, savings, credit, investment and risk, future value, regulations, tariffs, taxation, macroeconomics, recession, inflation, public goods, a public sector, consumerism, economic growth, economic specialization, market concentration and extensive trade. All these concepts would be foreign to hunter/gatherers but they are integral to an economy above the natural limit – they are automatically generated by the more complex economy. What started as a simple, direct system of supply and demand has evolved to be more complex, with more parts and complex dynamics within it. It is no longer a pure free-market economy – with its complexity, it cannot be. Microeconomic analysis is still relevant in a complex economy, but macroeconomic theory is necessary to attempt to understand the complexities of an economy above the natural limit.

Population and Technology and Growth

Danish economist Ester Boserup wrote about the interactions between population size and technological change. She recognized that two approaches were used by humans to harvest resources. One was to adapt the density of a local population to fit the abundance of the local environment through migration and management of the size of the population. Hunter/gatherers approached harvesting resources in this manner. The second system adapted productivity of the local environment to fit the demands of the population – using technology to modify the land to grow what the population required (Boserup 1981, 31).

Humans have several different systems of harvesting resources, gradated from "extensive" (harvesting a wide range of foods and food groups from the environment) to "intensive" (harvesting two or more crops from a plot per year). The systems Boserup identified are gathering (the most extensive), forest-fallow, bush-fallow, short-fallow, annual cropping and multicropping (the most intensive). Hunter/gatherers practice the most extensive system of gathering, while populations harvesting the most they can from a field in a season practice multicropping (Boserup 1981, 18-19; Price 1995, 3-4).

Boserup correlated the system of food harvesting used in a region with population density. She found that as a group increased in density, they opted for more intensive food gathering systems – they used more active manipulation to grow more than would have been possible from unmanaged land. This led her to propose that a denser population stimulated the generation of agricultural technology.

Anthropologist Mark Nathan Cohen came to a similar conclusion. In assessing why the Neolithic Revolution came to be at all, he wrote that population size was the driving force. He observed that all of the regional populations around the world that went from living as hunter/gatherers to developing agriculture at the Neolithic did so independently of each other, implying that something internal to each population was driving them to a similar point. Cohen believes that population size was driving those populations to agriculture (Cohen 1977, 5-14).

Much of what Ester Boserup wrote about focused on the development of agricultural technology as population changed, but she also noted that other, non-agricultural areas of technology developed when population densities increased. In areas of high density, infrastructure emerged from the need to move food from rural production areas to urban processing and consumption centers.

A main premise of this book is extremely simple: humans are a unique species on earth, able to create complex technology, and with this ability have uniquely been able to go past the natural upper limit defined by biological systems of the earth. Our technology allows the population to grow large and our large population prompts technology to keep evolving. This understanding is entirely consistent with Boserup and Cohen, who endorse the concept that population growth generates technological evolution. Evolving technology stimulates population growth and technology evolves because our population grows. This dynamic emerged at the Neolithic and we have been living these same dynamics for 10,500 years.

Our economy, a function of our population and technology, grows in parallel with the growth of both. Growth not only in size, but in complexity. Of necessity, our technology gets more complex as it evolves, making our economy and our culture more complex. Complex technologies, a complex economy and a complex culture are inevitable when our population is above the natural upper limit.

Chapter 14

Humans As a Biological Species

Biological life on earth relies on daily solar input for energy (Keeton 1972, 636). It starts when light from our sun strikes algae and the leaves of plants. Chloroplasts are energized to photosynthesize and create sugars and oxygen from carbon dioxide and water. When herbivores consume plants they are harvesting the energy plants gained from the sun, and use it to fuel their own bodies. Carnivores do not consume plants but gain the energy they need by harvesting herbivores.

This system, where daily solar energy is harnessed through photosynthesis to support the vast majority of life on earth, has been going on for millions of years. Solar input is foundational to ecosystems; photosynthetic organisms absorb energy from sunlight on a daily basis and all other life forms are supported by that first step. The finite supply of daily solar input limits plant growth. Every plant species requires different amounts of sunlight, at different wavelengths; without it, they would be weakened or perish. Reduced sunlight reaching the surface of the earth due to ash from major volcanos and debris from asteroid impacts resulted in significant die-offs of animals, whose plant food sources died for lack of sunlight.

As discussed earlier, the role of limiting factors within a biological system is important. Limiting factors control the size of the population of every species, disallowing over-consumption of resources by a species or the generation of an unhealthy amount of by-products from any technology they create. Since the lives of organisms in an ecosystem are connected to other species, by maintaining constrained population sizes, no one species has an undue impact on those other species. The relationships each species has evolved with others is maintained. While there are ups and downs of population sizes of all organisms, they tend to stay within a zone of compatibility with other species and maintain overall ecosystem health. The biological world is amazingly complex, but species in an ecosystem have evolved together and come to a balanced system.

However, when something from outside an evolved ecosystem enters in, the changes often cause some form of disruption. This is one reason an invasive species can wreak havoc on a regional ecosystem: the organisms of that ecosystem had not evolved with the invasive, so there was not an established, mutually beneficial set of relationships between them. In many cases, an invasive has an inherent advantage that allows it to grow with few constraints, generating a large population and an oversized impact on the ecosystem.

The effect of our technology is to generate undue influences on ecosystems. This happens in two ways, dynamics that have been discussed above: technology pushes back biologic limiting factors, allowing our population to grow beyond the natural upper limit, allowing the depletion of resources; and technology generates by-products that are pollution, in our case some extremely toxic, long-lasting and widespread pollution.

The technology we create and apply, and the influence it has, comes from outside the natural ecosystem we live in. Natural biologic systems are based on daily solar input, and this level of energy infusion allows a certain level of productivity, maintaining a population in balance with other organisms in the ecosystem. Our extensive use of fossil fuels to energize our technology allows us to be enormously more productive than daily solar energy would allow, and that productivity translates in part into population growth. But that productivity and our oversized population come at a cost: pollution and resource depletion. They are inevitable when elements from outside a natural ecosystem are applied to prompt growth within that ecosystem that is higher than the natural upper limit.

In a similar way, our outsized population generates a disharmonic effect on how we relate to each other. The natural history of every species is unique, and humans are wired to relate to each other in particularly sized groups. Anthropologist Robin Dunbar has done research that indicates that groups of three sizes tend to be stable and socially optimal: groups of 50 people, 150 people and 500 people. Groups in sizes other than these tend to be less functional (Dunbar 2017). Anthropologists tell us that hunter/gatherers tend to maintain groups of 50 people or fewer, and that larger groups tend to split. Hunter/gatherers were following the internal guidance of our genetics in searching for an optimum group size.

At the Neolithic, when regional populations grew but congregated in settlements and then cities, those population centers were larger than optimum for human wiring. With continued population growth, cities continued to increase in size. Without being able to know specifically what generates the psychological impetus to find harmony in particular group sizes, large and dense populations are difficult for humans to maintain healthy relationships with each other. In this way, large and growing populations bring with them disharmonious cultures. Technology and our large population have taken people out of their cultural comfort zone.

Chapter 15

The Human Population Growth Curve

A stereotypical population growth curve for a non-human organism would show that a population fluctuates over time based on the changing constraints of limiting factors. The maximum size of the population over a long period is its natural upper limit, reached at a point when limiting factors were optimum for growth.

For an example of these dynamics, we can refer back to the graph of the population of ruffed grouse (*Bonasa umbellus*) in Minnesota from 1982 to 2018 and see the growth and contraction of the population size over time and the natural upper limit. Population growth is a function of core, fundamental aspects inherent in biological systems: natural selection, survival of the fittest, survival as a primary drive, competition for resources and fertility rate of females, but the main influence on the growth and contraction of populations is limiting factors. This population of Minnesota ruffed grouse is subject to predation pressure, to disease, and to scarcity of the resources it needs to survive, limiting factors that both support and constrain population growth.

In this ruffed grouse example, population varies in response to the variations in limiting factor variables. The growth curve for ruffed grouse embodies changes that are more or less typical of populations – not specifically the rise or fall at any given time, but the concept that populations rise and fall based on different environmental variables, and that population size stays below an upper limit.

A graph of the growth in the human population from 10,000 BCE until 1990 CE takes a different shape, shown in Figure 15-1:

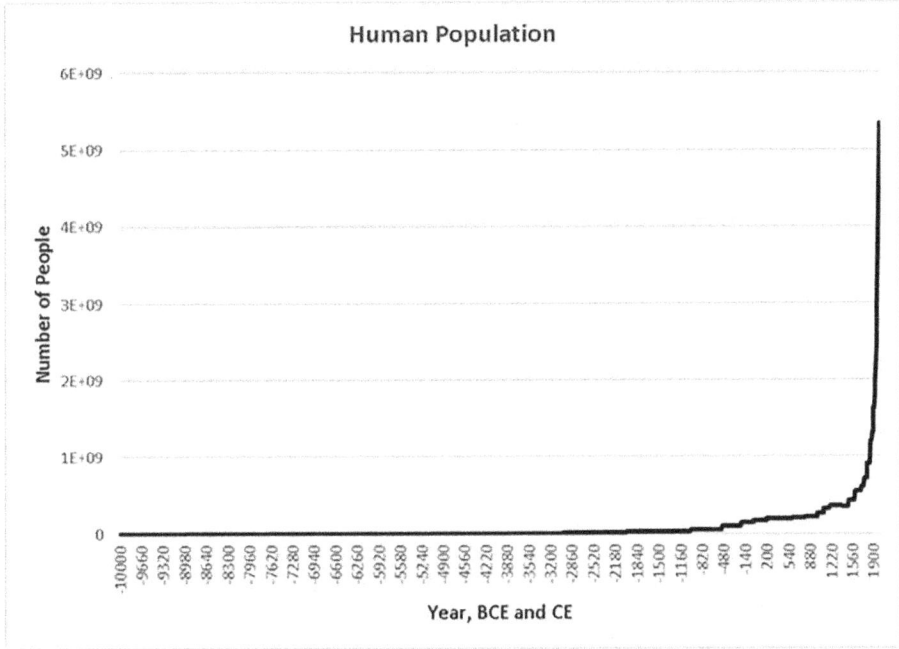

Figure 15-1 Data Source: Michael Kremer (1993)

Biologically, the difference between the example of the roughed grouse growth curve and the human growth curve is striking, and gives enormous pause. Humans are the only species in the world whose population grows in this trend, growing very slowly over a long period of time, then making a sharp change and growing very quickly over about 300 years. With this unique population growth curve, humans must have unique ways to affect the impact of the limiting factors on us. What does it mean that the human population growth curve follows a unique trend?

The Shape of Our Growth Curve

Homo sapiens diverged from *Homo erectus* about 300,000 years ago in Africa. We do not have finely granulated population data for humans between the beginning of our species and 1700 CE, but it is likely that regional populations vacillated up and down in response to limiting factor impact. Over that period, the population grew slowly. Graphically, our growth curve is close to horizontal.

Starting around 1700 CE, the human population growth curve started to change from a typical growth curve. Instead of expanding and contracting within a slow growth trend it started to grow at a sharp rate, and the growth rate increased with time. From about 1700 CE to 1965 CE, our global population grew at an increasing rate. Our curve turns from being a slowly climbing horizontal line between 10,000 BCE and 1700 CE, to an almost vertical line. From 1965 CE to now, our population has been growing, but at a slowing rate.

Between the Neolithic and 1700 CE

Humans shifted to agriculture at the Neolithic, adopting technology that allowed our population to grow. A question comes up: given the premise that technology allows our population to grow beyond the natural upper limit and keep growing, why did our population not surge in size at 10,500 BCE, in a pattern of growth we begin to see around 1700?

It looks like what happened is that it took just under 10,500 years of population growth and technological evolution to get to a critical mass of technologies that allowed our population growth to surge. After we started on the path of creating and applying technology, population and technology growth was incremental and relatively slow, but starting about 1700 and for the next century many technologies came together to reduce the incidence of disease, provide food, transport products quickly and efficiently and harness concentrated energy sources. Many of the technological changes started in England, and spread to Continental Europe and the United States over time.

Before about 1700, resource pressure in England and Europe had been increasing. Cities were getting larger as people migrated to them from the country, generating a need to get more food from fields to city and necessitating technology to deal with large volumes of human waste. Some countries were depleting wood supplies – wood was used for fuel to heat homes and for metal smelting – generating a look at how to mine coal more effectively. Advances in steam engine technology in England starting in the early 1700s were used to power pumps to draw water out of coalmines, allowing increased coal extraction. Metallurgy technology improved throughout the 18th century, especially efficiencies in producing iron, providing the basis for casting iron for machinery and later for making electric motors and external and internal combustion engines. Agriculture productivity increased significantly starting in the mid-1700s in England, supporting the growing population and prompting migration from rural areas to urban.

Three Observations

Three aspects about the sharp upturn of our growth curve are worth noting. First, rapid, sustained population growth is unique in the natural world. No other organism has a virtually vertical growth curve over a long period. Second, the change starting about 1700 from slow to rapid growth is the start of what's now recognized as a demographic transition, and is in large part a function of our applied technology reducing the incidence of disease in our population and improving our ability to grow food. These technologies decreased the mortality rate of our population simultaneous with the fertility rate remaining high, causing the population to grow quickly (Freedman 1974, 31).

Chapter 16

New Points on the Growth Curve

The third aspect to note about our growth curve is fascinating and significant. It involves several threads.

Anthropology informs us about the characteristics of hunter/gatherer populations. They harvested the resources they needed from the environment without overharvesting, and they developed a cultural framework that generated stability and harmony. These dynamics that defined their relationship with the natural world and their relationships with each other were effective in allowing our hunter/gatherer ancestors to survive for thousands of years.

By adopting agriculture at the Neolithic and initiating the process of population growth and technological evolution, we changed the dynamics of our relationships with the natural world and between people. Our larger population took us to a new spot on the growth curve, and our behavior changed in reaction to the new population dynamics.

At the Neolithic, the human population grew above the natural upper limit when we created and applied agricultural technology that pushed back the constraints of resource supply. This process, once started, progressed incrementally, growing and getting more complex. As we created and applied a technology to relieve a constraint, our population grew, but we immediately came against a new, different constraint, calling for either improving the original technology or creating and applying a new technology. Technology evolved and prompted our population to grow, and the growing population prompted technological evolution. Technologies build off those that came before, solving new problems, getting more complex and continuing to open the door to population growth – an incremental, sequential progression.

Our behavior in a small population, such as a clan size of 50 people or less, is different relative to large populations; beyond a certain size, complexity emerges, and with complexity different behavior (Simmel 1902, 1-3; Carneiro 1986; Dunbar 2017). When we shifted to agriculture, we congregated in settlements, built permanent shelters, lost a broad and deep connection with the land, gained deep knowledge of specific aspects of the land, political and economic hierarchy developed, inequality arose, and we started to develop specific areas of production in our economy. With time, cities formed and enlarged, generating governments. In a larger population, where no individual could have a close, personal relationship with every other person in the population, anonymity and distance allowed the less desirable aspects of our nature to emerge (Wirth 1938, 10-14). Conflict and dissention could be less constrained, go unresolved and have larger impacts. These behaviors developed as a function of a population larger than the natural upper limit.

The third important aspect of our growth curve is that after growing our population and breaching the natural upper limit, our behavior changed in reaction to being in larger groups, and continued to change as the population continued to grow. Because our population was and is growing, every point on the growth curve is new, and our collective behavior changes in response to the changing population dynamics at each new point. Because we are uniquely able to create complex technology and keep it evolving, and because agriculture is scalable, since the Neolithic Revolution our population has continued to grow and technology has continued to evolve, and our culture and the economy continued to grow and get more complex. In effect, each new point on our growth curve was a new environment, and our behavior changed in reaction to the new dynamics at each new point. We reacted in ways that are wired by our biological makeup, but in ways that are unpredictable because we have never been faced with the same cultural, economic and technological dynamics that are unique on each new point on the population growth curve. Our population has been doing this dance for 10,500 years.

Numerous Points Make Trends

If we take a macro-level view of some patterns of our lifestyle since the Neolithic, what elements show up? Our population is growing and has been for a long time. Each new population size puts the human species on a different point of the population growth curve and each new point is new territory for us. We react to new territory differently and changes must happen. What are some of the trends we are generating by our population growth? What do these trends tell us?

Economic Development

"Development" is the process a state or an economy goes through over time as technology evolves and as technology is absorbed into that economy. From experience we know that development evolves through phases: economies go from being agricultural to industrial and then to service economies (Shelp 1981, 13, 68). Starting in the United Kingdom in about 1750, an economy industrialized for the first time, marking the arrival of the Industrial Revolution. Trends of population growth, an increase in available food, the availability of inexpensive coal, the advancement of iron production technology causing the cost of iron to decrease, the production of functional steam engines and advancements in transportation all came together to allow the rapid advancement of industrialization (McNeill 2003, 230-235). Industrialization later spread to other regions of Europe and the United States, from about 1830-1850 (Kjeldsen-Kragh 2007, 17-24). Since those first shifts from agricultural to industrial economies, development has taken hold in different countries at different times. Currently, some countries are considered to be developed and some are considered to be developing, depending on the degree to which their economy has absorbed technology.

As an industrial economy matures, industry as a percentage of the economy tends to decline, but at the same time the service sector of that economy tends to grow. An economy is defined as a service economy when half or more of its labor force is employed in the service sector or when more than half of its Gross Domestic Product is attributed to service activities. The economy of the United States was the first to be a service economy, starting in 1940 (Shelp 1981, 1-9, 13-15, 22). Other economies followed. By the late 1970s, the Netherlands, Norway, Germany and Japan had service economies. Currently, 173 nations out of 220 listed have an economy whose GDP is greater than 50 percent service-based (Wikipedia).

What this trend in development points to is that economies evolve; more to the point, that technology, population size and economies all grow and evolve together. When assessed from a macro point of view, they are all connected: technology allows the population to grow, population growth prompts continued technological evolution, and the economy is a function of growth and evolution of both population and technology. Technology evolves into different realms as the growing population points out the places there is resistance to growth, the economy grows as more economic transactions transpire in a consistently changing marketplace and more people are making those transactions and expecting more growth. And all – technology, the economy and human culture – get more complex as they grow and evolve.

Science

There are many different ways to define "science." For the purposes of this book, a general definition is helpful: "Science is concerned with the material universe, seeking to discover facts about it and to fit those facts into conceptual schemes, called theories or laws, that will clarify relations among them" (Keeton 1972, 2).

People were studying the natural world as a way to benefit their lives before those behaviors qualified as formal science. People in Mesopotamia domesticated the dog, emmer wheat and barley by about 9000 BCE. By about 7000 BCE Mesopotamians were using techniques that would grow into systems of numeration and writing; by 3000 BCE Egyptians and Sumerians were using numbering systems to track agricultural output. By about 4000 BCE Egyptians were using a 365-day calendar. Egyptians were smelting bronze by about 3000 BCE, and by 1400 BCE Hittites were smelting iron. Chinese were making astronomical observations by about 2200 BCE. The Vietnamese were using bronze ploughs by about 1550 BCE (Hellemans 1988, 4-13).

Humans have consistently sought to understand the world around them, to be able to harvest and manipulate resources around them – hunter/gatherers and agriculturalists both have a stake in knowing what resources in their environment will help them survive. With agriculture, humans still relied on natural resources but the focus shifted from what resources were available naturally to how productivity could be increased by using technology. Humans have come to structured, formal science recently, but before that point, we were keen observers of the natural world.

While there is no clearly defined beginning to the study of science, the Scientific Revolution is said to have started in 1543 with the publication in Poland of *On the Revolutions of the Celestial Spheres*, Nicolaus Copernicus' treatise stating that the earth and other planets revolved around the sun, contrary to the belief in a geo-centric universe that predominated. In 1609, Galileo Galilei constructed a functional telescope that he trained on the earth's moon and our solar system's planets and made observations that effectively confirmed Copernicus' helio-centric model. Isaac Newton's publication in 1687 of *Philosophiae naturalis principia mathematica* laid a mathematical foundation valuable for the field of physics (Hellemans 1988, 90-146). An enormous number of observations of the natural world have been made since the Scientific Revolution in the fields of mathematics, physics, astronomy, biology and chemistry.

Broadly, the Scientific Revolution and the establishment of the empirical method to perform science was a shift from intuitive, informal observations of the natural world to structured observations that could be tested, measured and communicated, with the aim to learn about the natural world and use that knowledge to help us survive. The shift at the Neolithic to agriculture led to people applying those astute observations of the natural world to create technologies that would support the growth of their population.

There are different ways to assess trends that science and scientific observations have taken over the long-term. One way is to look at trends of basic and applied science. 'Basic science' or 'basic research' refers to scientific observations performed with the goal to discern fundamental laws and dynamics of the natural world. Vannevar Bush described it like this: "Basic research is performed without thought of practical ends. It results in general knowledge and an understanding of nature and its laws." (Bush 1945, 18) In a different manner than basic science, 'applied science' or 'applied research' is scientific work carried out with the goal to solve specific problems or result in a particular technology. Bush's report described applied research: "This general knowledge [derived from basic science] provides the means of answering a large number of important practical problems, though it may not give a complete specific answer to any of them. The function of applied research is to provide such complete answers" (Bush 1945, 18).

The observations of early scientists were for the most part basic science. When Copernicus, Galileo and Johannes Kepler made observations on the rotation of the earth and planets around the sun, they were seeking to understand the basic, underlying dynamics of planetary motion. In writing his book *On the Origin of Species* and in proposing a mechanism for species evolution, Charles Darwin was working to make sense of some fundamental organism dynamics he had observed in the natural world. Measuring the speed of light, observing the effect of gravity on objects, discovering chemical elements, observing the effect of a vacuum on falling objects, exploring the ability of heat to kill microbes – these are all observations of fundamental dynamics of the natural world: basic science.

Applied science works with laws and dynamics of the natural world that basic science has observed; once a basic law of nature is understood, it may be applied to create something useful. Examples include working to derive new medicines to attack specific diseases, designing electricity-generating power plants for use with energy-dense fuel, creating microscopes and telescopes with the understanding that glass lenses bend light and using magnetism to create a tool that images human bodies. The application of science to specific ends resulted in these technologies.

Basic science supplies the understanding of natural laws that supports applied science. In a similar way, applied science can expand or improve a particular technology. For example, after a basic science observation that some microbes attack other microbes, applied science developed antibiotic medicines in the 1940s to fight disease in people. With that opening, applied science developed more classes of antibiotics as the medical technology world saw the potential of the concept that basic science had observed.

Another trend in the sciences is that they have made observations at finer and finer detail as their work has progressed. Initial observations within a field tend to be of a macro-view (for example, that genetic characteristics are inheritable), and as those observations are understood, absorbed and worked with by the scientific community, observations continue at a finer and finer scale (genetic characteristics are transmitted by genes; genes are composed of nucleotides; nucleotides are composed of specific molecules). Another way to see this trend is that scientific observations have gained in complexity over time. The finer and more nuanced scientific observations become, the more science observes the complexity of the natural world.

The trend of scientific observations evolving from broad to fine is consistent with the evolution of technology. Initial observations of the natural world were without technology, but even simple technologies allow deeper insight beyond what the human body can sense. A curved piece of glass allows a view to a distant planet or into the workings of a microbe. The evolution of technology evolves in parallel with the increasing depth of understanding we gain by using technology. Technology can be complex, with many parts combined to make a tool that allows an insight into a natural system; and technology can allow a finer and deeper insight into the systems of the natural world as tools and understanding grow more complex.

Over time, scientific observations have gone more deeply into finer workings of the natural world and gotten more complex. These changes are a function of population growth and technological evolution – science is integral with the human population and its interest in understanding the natural world and is integral with the interest of technologists seeking more insights they can use to create technology.

Technology

Every species is a biological experiment, a new design evolved from a previous one and tested against environmental conditions. If the new species lives, the design is successful. If the species goes extinct, we can say that it was not adapted to its environment. Humans evolved with a unique combination of physiological traits, giving us the ability to create advanced technology. Our applied technology allows us unique ways to resist the impacts of limiting factors, with the goal of survival.

Other species create and apply technology, but only simple technology. They can develop their technology to a certain point, but are limited by their physiology and natural history to make it complex. Humans are unique in that we can take the concept of 'technology' as something that allows our population to grow and makes our lives easier, and keep it evolving far beyond simple. As Vannevar Bush implied, there probably is not a limit to what we can envision or create (Bush 1945).

And look at what that does: by pushing back the constraints that are inherent in the biological world, human's applied technology allows our population to grow beyond natural biological system constraints, beyond the natural upper limit. This is an enormously significant step. With our unique abilities we are able to develop new technology to push back against constraints that appear. By being able to continue this dynamic, we generate a population growth curve that is unique in the natural world. And on that curve, our human nature reacts to each new point of a larger population, including the dynamics of additional complexity, more crowded conditions, and changing patterns of transportation and resource acquisition.

Taking a brief look at the history of our technology, the earliest archaeological evidence we have of hominin technology is stone tool construction and use from about 3.3 million years ago, possibly by the species *Australopithecus afarensis* (McPherron 2010; Harmand 2015). Since those rough tools fashioned over three million years ago, our ancestors used stone to make many tools, refining their ability to shape stone over time. Many of the tools developed by hunter/gatherers and for agriculture were stone, including sickles, hoes, axes, adzes, scrapers and points for spears and arrows. The predominance of stone for the technologies of this era is why it is termed the Stone Age.

Along with stone, our ancestors likely used other natural resources – bone, wood, antler, plant fiber, wood, leather – to create tools, including fishing hooks and fishnets, harpoons, needles for sewing and clothing. They also used fire: possibly to alter their landscape, and by about 400,000 years ago to cook food (Wrangham 2010; Attwell 2015, 1-6).

These technologies of the Stone Age, up to the Bronze Age, were all relatively simple: they were made of minimally processed natural resources, for the most part using muscles to create the technology and to use them. Wind energy was harnessed for sailing boats, probably about 3000 BCE.

At the Neolithic Revolution, starting approximately 10,500 years ago, humans shifted from a hunter/gatherer lifestyle to agriculture and in doing so committed to relying on technology to support their population. This started humans on a path that would make technologies more complex. They developed agricultural techniques to manage food production, created tools to assist in agricultural production and to store harvests, domesticated plants and animals, created permanent dwellings, manipulated water flow and developed irrigation systems.

About 3000 BCE in the region now referred to as the Middle East, people started mining and smelting copper, tin and arsenic, the raw materials to create bronze. This marked the beginning of the Bronze Age. By about 1000 BCE humans were creating iron. The creation and use of metals prompted new technologies as people explored the properties and values of these new materials. By about 1700 CE, the use of metals to create machines and later, electric motors, soon thereafter led to the Industrial Revolution.

Starting in about 1940 CE developed economies evolved from being industrial to information economies, where information and data are the primary focus. Information economies develop and use a wide array of electronic tools to store and manipulate information for the purposes of industrial production but also for the production of services (Shelp 1981, 1-9). An information economy is sometimes referred to as a knowledge economy, a service economy, or post-industrial.

Technology is a fundamental element in our lives, has been for 10,500 years, and we have kept it evolving since committing to it at the Neolithic. Given the significance of this dynamic, what technology trends have we generated?

There are several trends embedded in our history of technology. One is that human technology was initially very simple, comprised of a few tools constructed by our hunter/gatherer ancestors from minimally processed natural resources, and that over time our technology has gotten more complex. No longer based on simple modifications of natural resources, our tools are now created through multi-step processes of harvesting and manipulating resources to produce highly refined products. Technology has tended to get more complex as it has developed through different, extraordinarily complex stages. Our current technologies are made of highly refined natural resources, many made up of a vast number of parts, and many people of the global population are influenced by complex technology.

Another trend is that much of our technology is geared toward processing large volumes of raw materials. Hunter/gatherers created and applied technology to reduce the work it took to harvest the resources they sought, but they harvested for immediate need, not beyond. Hunter/gatherers knew their environment and were capable of finding the resources they needed for their small population and storage of food was rarely possible, so they would gather and hunt resources as needed and consume relatively quickly. Agriculturalist needs are different and much of our technologies involve harvesting huge quantities of resources. Our consumption of mineral ores to make metals, fossil fuels to provide energy that powers our technologies, trees to fabricate paper and construction products, immense quantities of water for industrial processes and residential consumption, food production on huge farms using enormous quantities of herbicides and fertilizers, and the fabrication of huge quantities of building materials are examples of this (Steffan 2005, 82-89, 131-134). We consume without regard for environmental consequences; our interest in harvesting what we need to support the large population over-rides any sense of harvesting and consuming within the limits of what the biological world can produce within the daily solar input system.

Another technological trend is that our economy is now based on using energy-dense fuels to energize our technologies. For the majority of our history humans performed the work it takes to survive by using our muscles. As hunter/gatherers and as agriculturalists, work was performed by people until about 3000 BCE when we started to harness energy from wind, and later, flowing water. About 1700, we turned to the fossil fuel of coal for an energy source at a time when we were consuming trees faster than they could regenerate, and we have not looked back since. The energy of fossil fuels powered the Industrial Revolution and they continue to energize our current economies of the globe (Cleveland 2014, 340).

What is the significance of these trends, of technological complexity, of technology that has become highly refined, of the need to harvest huge volumes of raw materials, and needing to utilize energy-dense fuels to energize the technologies that support us? What do these trends tell us?

Because of each of these trends, we are suffering long-term and deleterious pollution, severe resource depletion and a culture that is difficult to maintain in harmony. The size of our population and the technological complexity we have developed by pushing beyond the natural upper limit result in the pollution, resource depletion and cultural dysfunction we experience now. These trends give us insight into dynamics that are affecting us.

Culture

Culture is difficult to define – the concept is a broad and amorphous one. Intuitively, culture encompasses the elements of social interaction generated by human behavior within a population. The work and recreation that take place, the habits and norms and traditions that form and change, the rules that define relationships between people, the constructed environment, the political structure, the type of economy, the religions within the region — pretty much everything people do within the population — are elements of culture. And the ecosystem affects a culture; people respond to the land and climate as they perform the work of living in a region. Cultures of the arctic differ from those in temperate climates, in part because of the ecosystem.

Given this inclusivity, cultures differ from region to region and change over time. Technology changes, specific religions become more popular or decline, music styles evolve, long-term weather patterns change; culture responds to all these changes. People's behavior affects these patterns, and the patterns affect people's behavior.

Hunter/gatherer cultures were simple and the rules clans lived by were effectively the result of cultural natural selection – when individuals of a group work together using these guidelines, survival was more likely for each and for the clan. Hunter/gatherers functioned in ways that maintained close relationships with others in their clan and maintained healthy relationships between the clan and their ecosystem.

At the Neolithic Revolution, in committing to population growth and technology, we severed ourselves from hunter/gatherer principles that kept us functional and stable for thousands of years. The shift to agriculture required changing to a regime of larger populations, fewer close connections within the population, more personal expressions of negative emotions and egoism, hierarchy and less ability to generate mutual support within the group. Living in larger, denser settlements and using advanced technology changed how we live, work, relate to each other and relate to the environment. Large populations and advanced technology have the effect of creating distance between people and less of a connection between a population and their ecosystem.

When we live in large populations, we can no longer maintain close relationships with all other people around us. As the anthropologist Robin Dunbar observed, our biological wiring places a limit on our capacity to relate closely to other individuals in large groups (Dunbar 2017, 208-211). Small communities develop organically within larger populations, but those outside of the small groups are effectively removed from relating closely to individuals within the group. Biologically, people outside the small community are of less benefit for its survival, and pose a threat to the resources of the small community. These dynamics are the same as those hunter/gatherers experienced: maintain harmony within the clan to promote survival, but defend against those outside the clan.

Contrary to the close relationships in hunter/gatherer clans, diminished connections to others in a larger population fosters a climate where negative emotions can bubble up and there are fewer motivations to conform to cultural guidelines. In larger populations, individuals no long rely on each other for support, and awareness of this means incivility and conflicts are less constrained. Group cohesion and harmony are more likely to suffer (Wirth 1938, 1).

Technology affects how people relate to their environment. We have made an incredible number of tools that replace people to do work and energized them with dense energy sources external to our bodies. As a result, our culture is one of productivity, but we are also removed from the connection that comes with doing work with our bodies. Technology has changed how we travel, grow our food, create products, move products, cook our food, build and heat our homes. In the past, those activities involved a close understanding of resources in the environment. The very nature of technology is that it is an intermediate step between an individual and a personal experience in the environment; and by being intermediate, it precludes a direct experience.

The amount of technology in our global culture today is tremendous, and permeates all aspects of our lives. These added elements make our culture enormously more complex. Relationships between people are diminished along with connections to the natural world.

The shift to agriculture brings a secondary dynamic. With population growth and technological evolution, not only do we lose the ability to maintain the policies of hunter/gatherers, but also the ability to adhere to the principles those policies were based on and that kept clans intact over long periods of time: harmony and stability within the clan. Hunter/gatherer policies defining acceptable behavior within the clan were derived from the principles of seeking harmony and stability, which served their interest in surviving for the long-term. Hunter/gatherer policies benefitted both individuals and the clan: the goal is survival, and their policies shifted the odds toward being able to survive. But living in large, dense populations and applying complex technology meant that the policies and principles of hunter/gatherers were difficult, perhaps impossible, to maintain.

In a hunter/gatherer population, the group engages in harvest together and mutually allocates the rewards. With agriculture, the population produces needed resources through specialized and hierarchical work. Some people own the land and technologies to create products, some work as laborers, some work in niches creating technology, some work in political niches and gain resources through the labor of others. The acquisition of resources no longer relies on harmony within the group, but on population and technological growth, which demands planning, structured work and investment. And resource allocation is not necessarily based on the principle of sharing and reciprocity but through hierarchy; those with more status or power or investment may acquire a larger slice of the resources. Consequently, mutual group effort is not needed or even helpful for survival of an individual or the population. The risk of death or ostracization from not helping the community, a driving force for hunter/gatherers, is minimized in agricultural cultures.

Culture, being a function of many elements of human behavior gets more complex as any of these elements get more complex, which happens as a population grows above the natural upper limit and as technology evolves. Where a hunter/gatherer culture has relatively few cultural elements to manage – population size, maintaining clan harmony – and is consequently stable over the long-term, the culture of an agricultural population inherently has more elements embedded in it and is changing constantly as it grows, making it more complex and difficult to manage.

This discussion examined the cultural changes that took place when humans shifted from being hunter/gatherers to agriculturalists, but with continual population growth and technological evolution, culture kept, and keeps, changing and becoming more complex. As there are continually more people interacting, and more technology that people interact with, the economy continues to create specialized niches. We experienced a significant shift that made culture more complex starting about 1750 as England began to industrialize: many people in rural areas migrated to cities, technology began to mechanize labor dramatically, we began to understand how disease was transmitted and worked on ways to slow disease spread, and population was growing.

Industrialization expanded to western Europe and North America and then to other countries. Every new point of growth in the population brings a new point on the growth curve – as the population grows, technology evolves and gets more complex, and the culture gains in complexity. We are experiencing uncharted cultural territory on the growth curve constantly.

Growth Curve Summary

Our culture and our economies have been evolving ever since humans took the step of controlling the production of our food supply. Responding to environmental constraints that kept our population from growing or by simply seeking to make our lives easier, we have incrementally created and evolved our technologies. Our culture and economy have changed, evolved, morphed at every point on our growth curve in response to our growing population, and the trends these points form are not by chance. Our reactions are a function of innate *Homo sapiens* wiring as we respond to living in larger populations and the dynamics they bring.

The trends of the changes we make in response to living in larger populations are all around us. Trends visible from a macro perspective are described above, but we are completely submersed in these diverse trends. Our large, growing population and evolving technology explains the trend of rising expectations of education for employment, the evolution of communication technology, the growth in energy consumption, the increase in foreign travel by people from developed nations, the specific problems the sciences research, economic globalization, recessions and inflation, the need for central banks, the increasing complexity of the global financial industry, pandemics, species extinction at high rates, burning rivers, the "Green Revolution," dying languages, and more.

Each person in a population interfaces with their environment when they seek to gain a resource. Each person interfaces with other people in their population in the social dynamics of everyday life.

What those interactions look like is a function of a human nature that is wired into each individual. How someone reacts to interfacing with one other person in a population of two will be different from how they interface with 30 other people in their clan and those interactions will be different relative to interfacing with a population of 1 million.

This is an interesting dynamic for the human species. Our unique ability to create technology allowed us to launch on a path untrodden before in the biological world. With technology, our population grew larger than biological systems would naturally allow, and our large population generates behavior in us consistent with our biological makeup but unknown in populations smaller than the natural upper limit. Since our food supply is scalable and we have the ability to continually evolve our technology, the population not only grows beyond the upper limit, it keeps growing in a sequential, incremental manner. This is what we call "progress" and "improved quality of life." As our population has been growing, we have been continually reacting to living within the continually changing dynamics. A critical mass of technological expertise came together around 1750 that generated significant improvements in productivity and health that allowed our population to grow rapidly. We are in a unique situation.

Chapter 17

Summary

Humans are unique in the natural world, having evolved the physiology of opposable digits, a large brain and the ability to communicate complex concepts within the species. These traits allow us to envision, experiment with, create and apply complex technology. Our applied technology pushes back the impacts of biological limiting factors that would normally constrain the size of our population, and technology makes life easier for us. With limiting factor impacts reduced, our population has grown larger than the natural upper limit. A large population prompts technological evolution and technological evolution stimulates population growth. The side effects of our complex technology and our large population are pollution, resource depletion and a disharmonious culture.

Chapter 18

Conclusion

For hundreds of thousands of years, our hunter/gatherer ancestors accepted and worked with their environment and the resources it provided them. Like all other species, they knew how to gain what they needed from the natural world, they knew it would supply what they needed, and they knew to not over-harvest resources from their environment. Hunter/gatherers structured their culture with rules that maintained harmony and stability within their clan populations in spite of traits of our human nature that might otherwise prove divisive. In these ways hunter/gatherers maintained healthy, positive-sum relationships with their environments and with each other, relationships that could continue for the long-term (Lee 1999, 1, 392-393).

The Neolithic Revolution marked a significant departure from the hunter/gatherer lifestyle. Humans are extraordinarily adept problem-solvers, but in shifting to agriculture, we committed to two assumptions anathema to hunter/gatherers: that we need to grow our population and that more technology will be beneficial to us. These assumptions are constraints: in accepting them, humans lost sight of the bigger picture of our place in the natural world, a vision hunter/gatherers lived. The solutions we developed and applied based on those assumptions led, and continue to lead, humans away from healthy relationships with our environment and with each other. They are solutions for the short-term, not the long-term (Redman 1999, 91).

A distinct, sharp tension comes with our large population and the complex technology we create. We are driven to seek them, and in the short term they make our lives easier and longer and our population large and seemingly prosperous, but they bring two significant, existential problems. One is environmental: our technology and population growth are causing widespread and long-term pollution and resource depletion. Degradation of the resources we need to live threatens our long-term survival. The second problem is that while technological evolution and population growth feel like appropriate policies to pursue, with those come the generation of inequality and the expression of divisive behavior, resulting in cultural dysfunction. This dysfunction could easily erupt into widespread strife that could cause significant social trauma.

Our species has a consistent history of collapses of large, technological populations. Populations grew beyond the natural upper limit, were supported by technology, and collapsed during crisis. The risk of a population collapse now is real, in spite of any denial of that risk. That technology imparts downsides in different forms to humans is a recognized phenomenon (Chiles 2001, 1-8; Kates 1995, 1-4).

This tension of our human technology and our large population is exemplified by the recognition that the moment we set foot on the technological path we were trapped by it. We had to keep technology going or risk a population decline if technology was removed and there is tremendous tension in that risk. Humans have been living with that tension since the Neolithic. Currently, it shows up in the relatively minor economic shocks we feel. A company or an industry cannot be allowed to fail in a time of economic duress because we cannot allow jobs to be lost. Ebbs and flows happen constantly in the natural world: as limiting factors change, an organism's population size adjusts. For people, jobs equate to income and income allows families to eat. "Too big to fail" implies we are required to keep propping up a sector to keep jobs available and families fed for fear of collapse. Our large population lives with that fear every day.

We could benefit by looking to the natural world and understanding the biological system that all species, including ours, is part of and works with. We simply cannot get away from being integral with the biological world. Nature only and always applies the laws of physics, chemistry and biology, and resources are always in finite supply. We can deceive ourselves in thinking we are in control; but we are not. That self-deception leads us in a detrimental direction.

Biologist Edward O. Wilson spoke in a video about the difficulty we visit upon ourselves: "We are a dysfunctional species. Dysfunctional, and why? Because we have Paleolithic emotions, we have Medieval institutions, and on top of all of that, we have developed God-like technology. And that's a dangerous mix" (Wilson 2015). Some researchers in the field of evolutionary psychology bring a similar perspective to their work: the understanding that our brains are wired because of millions of years of natural selection in the natural environment, and those wirings drive our modern behavior even though the dynamics of modern life are dramatically different (Cosmides 1992, 5). We seek technology because from a natural selection perspective, it helps increase the odds of our survival. We are driven to reproduce because population growth is not typically a sure thing in the natural world and we are wired to seek it. But with our unique evolutionary abilities, advanced technology and a growing population are harmful for our long-term viability, not at all what we need to survive.

We would do well to take this message to heart. Every environmental and cultural problem we experience now is of our own making. Every one. No natural laws have been broken and all natural systems are working properly. The difficulties we experience are entirely consistent with laws of physics, chemistry and biology. If we generate radiation that disrupts DNA replication, resultant disease is entirely consistent with the laws of nature. If we overdraw water and people die for lack of water, that is biology. If nitrates consumed from a contaminated water supply prevent oxygen from being taken up by hemoglobin, the resultant illness is entirely consistent with natural life systems. If a group within a population is refused access to resources of the culture and become angry, resentful and alienated, that is the biology of our human nature being expressed. There should be no surprises at the difficulties we experience in a large population with complex technology. Nature doesn't care if we grow our population large and generate deleterious by-products that hurt us, and nature doesn't care if we generate traumatic cultural disharmony – nature only applies the laws of physics, chemistry and biology, without quarter.

We make technology because we are good at it. As skilled problem-solvers, we see what is holding us back from the population growth we desire and we have the ability to make stuff to resolve those constraints. Just like every other organism we are wired to do all we can to promote our survival, and technology is one thing we know to do. And it feels right to apply ourselves in this way: we are using our inherent skills to create technology that helps people live longer, can make us healthier, and make our lives easier. But because we are unique in our technological abilities, that path happens separately from the normal functioning and normal constraints of the natural world. We are responding to our psychological wiring that tells us to use every possible tool to survive and to grow our population to resist being taken over by neighbors, but too effectively. When we pump pollutants into the sink of the environment, when we draw resources at rates faster than they regenerate naturally, when we don't respond to systemic dysfunctional relationships between people, we're ignoring trends that can only come back to hurt us. Nature's laws cannot be broken or amended.

In the introduction, the questions posed asked about the drivers of environmental and cultural problems we are experiencing. Technology evolution and population growth and size are driving those problems. What is presented in this book is a model, summarized in chapter 17, a template of explanation for problems that we are experiencing. This model fits the observations of several fields of study and explains the way population and technology have the effect they do. The current track we are on is problematic. Getting off that track is the only chance we have to not fall into a societal collapse. Our large population and our use of technology are inevitably generating a risk of collapse.

What Can We Learn?

Vitally, there is an important message within this model that explains the unique place humans have in the world and how we have come to relate to the environment and relate to each other in the ways we do. From one perspective – assuming that we need to seek a growing population, that we need continued technological evolution, that we need to continue our economic expansion – the message in this model we are living is ultimately grim, depressing and bleak. If we truly believe we need to continue on the technological path we are on and that anything but that path is going backwards, then the model in this book is unacceptable.

But if we drop the assumptions that we need population and technological growth, absorb observations from the bigger picture and allow our perspective to shift, the content of this model becomes terrifically empowering. What is it that we, as a species, seek in the big-picture? If we seek a viable, healthy and harmonious population, with individuals able to access the resources needed to survive on a long-term basis, this model shows us that we lived that life at one time, and that the framework for how to live that life is still available to us. The natural world functions for the long-term – species are able to live and function healthily for thousands of years, hundreds of thousands of years, millions of years. All organisms live their lives in synchronicity with the natural world, harvesting and consuming the resources they need to survive and live healthy lives. *Homo sapiens* lived in that manner for a long time without generating pollution, without depleting resources and in populations in relative harmony.

A Healthy Environmental Dynamic

Just because we have the ability to create complex technology that causes us long-term problems does not mean we have to continue doing so. We can instead decide to live within the boundaries of the natural world, responding to what the laws of physics, chemistry and biology show us. We are intelligent enough to observe those laws, and intelligent enough to make a choice about the direction we want go.

Environmental dynamics of ecosystems show us that there is a distinct limit to what the biological world can produce based on the daily solar energy that can be harvested by primary producers. Daily solar input sets the upper boundary of biologic growth. This is a system that has been in place for millions of years and is the framework for all life forms. It is a system that has produced an astoundingly prolific and complex ecology of life forms, all interacting with each other.

To go beyond the biological boundary defined by daily solar input requires the addition of some form of energy from outside the system and the application of technology to harness it, but these inevitably cause some form of dysfunction. We have tried this and are experiencing the pollution and resource depletion that come with complex technology and a population larger than the natural upper limit. We have not found a technology that does not have an environmental cost to it; we have to assume it is impossible.

What this tells us is that we need to decrease our population size and reduce our creation and application of complex technology. When our population size is stable and can maintain itself for the long-term on resources grown from the daily input from the sun, and when our technology does not cause long-term or widespread pollution, we are living within the limits defined by the earth's biological systems.

A Healthy Cultural Dynamic

It appears that humans are predisposed toward behavior that causes divisions within clans and within populations. We have egos, individuals can be corrupted by accumulating power, individuals tend to be defensive and can become angry, we chose to be violent when threatened or think we're threatened – all these traits, when expressed, can result in divisions and dysfunctional dynamics between people within a culture.

But hunter/gatherers figured out how to avoid the culturally harmful effects of our human nature. Each of the pre-Neolithic Revolution populations applied similar rules to structure their cultures: equality of individuals; sharing and reciprocity of resources; small group size; communal resolve of disagreements; low-technology lifestyle; minimal accumulation of possessions. In spite of the tendencies of our human nature that if allowed to run rampant might cause divisions within a population, hunter/gatherers created a cultural template that by many anthropological accounts worked to maintain harmony and cohesiveness within their clans. It was a choice on their part and very likely a function of cultural natural selection: they all do better when they work together as a group and when this becomes evident, such behavior is selected for in a Darwinian way.

We have the same option now. If we seek a harmonious and cohesive culture that can function for the long-term, we can set policy to design the rules to favor the outcome we seek. The model expressed in this book, and the dynamics of the natural world, show us what is possible. Policies that design our culture are up to us.

What Do We Want?

We seek conflicting elements: we are wired to avoid pain and discomfort and to seek pleasure and ease; and yet the way we go about living those wirings is putting the quality of our future lives in doubt and is causing deep cultural discomfort. We simply cannot have short-term ease and luxury and a large population generated by complex technology simultaneous with long-term health and cultural stability. Because of our unique physiology we are uniquely able to create complex technology that generates pleasure and ease for the short-term. In the long-term those same technologies generate destructive elements that can destroy us. We cannot have it both ways. Ease for the short term inevitably results in dysfunction in the long-term.

This is an extraordinarily simple and extraordinarily difficult policy problem. If we are truly interested in a long-term existence as a species we need to change assumptions from what we currently are invested in: seeking population and economic growth and applying complex technology. Defensiveness, denial and reliance on the *status quo* will not help us. This is not a political issue. We have extraordinary abilities but we are using them to create short-term benefits, not longevity. We can continue to be shortsighted about the broader impacts of our actions, continue to care more about immediate ease than long-term benefits, and show that we do not have an interest in working for the long-term health of our progeny or our species. Or we can use our capacities to change behavior and live for the long-term.

Chapter 19

How Can We Move Forward?

Okay, so where does this leave us? This book provides an understanding of how people have come to experience the environmental and cultural problems they have, but the intent is to use this knowledge to improve our situation: to resolve our environmental problems and to resolve the disharmony within our culture. How can we move forward?

The model stated earlier is incredibly simple. It shows us that complex technology and a large population have a deleterious influence on the dynamics of human's relationships with the environment and with each other. This simplicity translates into an understanding of the actions we need to take to remediate the negative impacts we are having: we need to decrease our population size and stop using complex technology. If we could do these two things, problems of the environment and of our culture would dissipate.

We can learn from the hunter/gatherer lifestyle. It sets the bar at a sustainable level. Human life previous to 10,500 years ago was in synchrony with the environment and could have continued indefinitely. Hunter/gatherers knew how to exist in a harmonious relationship with the environment and in harmonious relationships within clans for the long-term. Humans did not pollute or deplete resources, and we created cultures that generated harmony. There was undoubtedly violence and warfare between clans, something that seems unavoidable given human nature.

How can we structure the steps we need to take to be a sustainable species?

Decrease the Population Size

Many in the academic, public and some religious sectors assume that population growth is a positive dynamic. That desire for growth may be a drive wired into our psyches. In the biological world it is rare that survival and population growth are assured; there is constant pressure from limiting factors that seeks to decrease the size of a population, so it makes sense that natural selection would wire organisms to seek growth. As well, the human impetus for population growth goes right back to the need to defend against large populations who might seek to overtake smaller populations and for the need for increasing labor supply to maintain a growing economy. That fear of being a small or shrinking population, relevant in our earlier history, is no longer reason for us to seek population growth – this is no longer the factor that creates political or military strength. Given our ability to grow our population far beyond what is sustainable for the long-term, we need to change our hardwired expectations for growth.

The people of some countries are in the process of reducing population growth. They are doing so not on the basis of seeking to reduce the negative impacts of our large population but as a function of the feedback they are getting from living in a large and complex population. Countries whose economies have evolved significantly are more developed and they tend to have lower birth rates: women living in developed economies recognize other opportunities for them in addition to being mothers, the need for large families to work at home diminishes and children become more expensive to raise. The growth rate of these countries is decreasing as a function of choices, conscious or sub-conscious, on the part of the population.

That individuals of a population chose to have fewer children is significant. The drive to have children comes from different sources – internal wiring, social pressure, religious beliefs – and the impetus is a healthy one. But individuals can make choices; if they look to the larger picture and recognize the problems that a large human population brings to the world, choosing not to have children is a rational and intelligent decision. For all other species than humans, and for humans who lived as hunter/gatherers, the environment defined the size of the population. Now, we need to consciously recognize that shrinking our population size and maintaining a small population are good for the long-term health of everyone.

National reproduction policies and limits have been legislated in the past by various countries, but it seems draconian to expect a world-wide policy could be set in place to limit population growth. At a personal level, individuals can have an effect, and their actions can drive national policies.

Even as the birth rate of the global population is decreasing, the global population is large and increasing. The difference between the rate of growth and the growth in overall size is important. Even when the growth rate falls to zero sometime around 2100 at an estimated 11.2 billion people, this size will be unsustainable. We need to bring the size of the global population down, and this can happen through individual choices. The birth rate of a couple, of a region, of a country are not set in stone, they are variable. Those rates are a function of different variables: interests of the females and of couples, the demands of the home, cultural dynamics, availability of resources to support life, and probably more. The important point is that if people are interested in maintaining the human species for the long-term in a healthy environment, we can make the choice to decrease our population to a size that is biologically sustainable.

Decreasing Application and Evolution of Technology

The other element that needs to change is our application of complex technology. There is simply no way to refine natural resources to a high degree to produce and apply complex technology and not have deleterious by-products from that process. To fabricate metals, to produce chemicals, to refine petroleum to make fuels and plastics, to support the systems that produce these technologies and that support the large population inevitably generates by-products that are polluting. That seems to be a physical reality, and unavoidable. We can wish it otherwise, but the limits of the physical world are real.

Decreasing the use of technology is necessary to stop producing the pollution and resource depletion that are harming us. All environmental problems we are concerned about are generated by our complex technologies. It is obvious that technology has changed our lives over the past 10,500 years and many believe the changes are for the better, but that technology is not necessary for humans to live comfortable lives. We lived for a long time without this evolved, complex technology, and we thrived.

Every item we create comes from natural resources that are harvested somewhere, transported elsewhere, fabricated and will need to be disposed of. Every step of that process has the potential to produce pollution and to deplete resources. Individuals can make choices to buy less stuff, and to buy and apply less technology. In consuming less, there will be less pollution.

It would be possible to apply regulations to limit the growth of technology and decrease consumption in the market, and this would have the effect of slowing or stopping pollution. But, a deeper fix would be for individuals of the population to decide to consume only what is really needed. Market demand, or lack of it, will drive a decrease in production on the supply side. If we all stopped using plastic, it would not be produced. If we stopped using fossil fuels, it would not cause pollution. If we stopped buying things we do not need to survive, unnecessary things would not be produced. It is that simple.

Energy of an Economy

An enormously significant dynamic of our contemporary economy is the use of fossil fuels. The injection of fossil fuels into the production processes around 1600 in England provided the energy that supported the Industrial Revolution, energy that was far more concentrated than previous energy sources of muscles, wind, water and wood. Without fossil fuels production couldn't have increased dramatically the way it did; the use of coal allowed the production and application of the steam engine, coal enabled an enormous growth in the production of iron and cast iron machinery, a rapid evolution of transportation technology, and in general a huge surge in economic growth. The population increased commensurately as the economy grew.

In recognizing this dynamic we can see that one of the most helpful practical steps individuals can take to help resolve our environmental and cultural problems is to stop using fossil fuels. They inject a concentrated energy into the economy that generates higher productivity but the environmental costs are huge and unavoidable. By stopping the consumption of natural gas, petroleum oil and coal, we would stop the release into the environment of CO_2, mercury, nitrogen and sulfur compounds, unrefined petroleum from spills, methane, persistent organic pollutants, fertilizers, herbicides, plastics, and more.

Nuclear power, while it does not generate the same by-products as fossil fuels, is harmful in a similar way. The wastes from using uranium and plutonium to generate power are incredibly toxic for an incredibly long time and we cannot avoid their harmful effects. The power of the atom is another highly concentrated energy source that is far beyond the system of daily solar input. There is no way for us to tap nuclear power for our long-term benefit.

At the moment, we are seeing an increase in the development and application of technology to harness solar, wind and water energy, and these are seen as "renewable," meant to supplant fossil energy. But this misses the point. In developing these technologies we are trying to maintain the system based on fossil fuels with alternative energy sources, which is not feasible. Alternative energy sources still need a vast number and volume of resources to work: steel, aluminum and copper, concrete, plastics, silicon, various elemental substances, heavy metals and water. Harvesting these materials to create alternative energy technology in the amounts needed to support our large population is environmentally problematic. The same issues emerge as in the creation and application of any complex technology: in the harvesting, refining, use and disposal of these technologies, pollution is inevitable.

These technologies may sound and feel good, but they continue the deleterious trends we observe. They do not address the long-term problems. The assumption they are based on is that they need to support a large population and they do not address the problems of pollution and resource depletion. These technologies are not based on understanding the larger biological system of which we are part.

The resolve for these problems is incredibly simple. Stop driving. Start walking. Heat houses to 55 degrees in the winter and turn off the air conditioning in the summer. Consume less electricity. Stop flying airplanes. Stop applying manufactured chemicals to agricultural fields. Stop the manufacture and use of all plastics. These steps are incredibly simple, but they would have a huge effect on reducing the environmental impact humans are having on the environment. The energy of fossil fuels, concentrated from plant and animal growth millions of years ago, is powerful for how it has changed our lives but it is also proven to be a stunningly deleterious product for our long-term health. We cannot function in an economy based on fossil fuels for the long-term.

As a species, we need to get to the point of producing and consuming on the basis of daily solar input. This is an extraordinarily simple concept. Growth and productivity from daily solar input is the dynamic that the biologic systems of the earth, and biologic sustainability, are founded on, and it is only in the past 400 or so years that we have incorporated fossil fuels into our economy in a significant way.

Coal was not a significant source of energy before about 1600, internal combustion automobiles were uncommon before 1900; the first synthetic plastic was developed in 1900; and before about 1940 all farms were organic, with little fossil fuel-based chemicals applied. We do not need fossil fuels to live; we can reduce our consumption of all fossil fuel products if we chose to. It is impossible to use fossil fuels without a steep environmental cost.

Net Effects of a Smaller Population and Less Technology

As the first part of this book describes, population growth and technology evolution are inextricably linked. The growth of one prompts the growth of the other, in a continuous cycle. In the reverse dynamic, by stopping the application of technology the population would have to decrease: technology provides the resources a large population needs to survive, so less technology means fewer resources and a smaller population.

Theoretically, decreasing our population would stop the evolution of technology and slow or discontinue the use of existing complex technology. But this is an uncertain dynamic. While population and technology support and prompt each other to grow and evolve, we don't really know what will happen when existing technology is unsupported by a decreasing population size. Could existing technology still be used? Would we try to continue technology evolution in spite of a shrinking population because we know how? This gets complex, at the intersection of human nature, human skills, acquired knowledge, the drive to survive and the application of intellectual understanding. It is uncertain. However, the bottom line is that we need to reduce the application of complex technology, whether by decreasing the population or by policy.

That we need to decrease the size of the population and stop the evolution of technology are radical proposals, and run counter to strongly accepted beliefs. But they are what we need to do. Our species simply cannot live sustainably with a large population and complex technology.

Cultural Dynamics

Culturally, remediation efforts need to emanate from the policy arena. Look to the rules that hunter/gatherers put in place: they were nomadic, they resolved disputes within the clan publicly with all clan member weighing in, they had no hierarchal structure and they shared and reciprocated resources within the clan. Nomadism stems from an environmental basis, so we can ignore it in this cultural discussion, but look at what the other three rules do for a culture: they minimize inequality within the clan.

Between any two people there are inequalities. One has better eyesight than the other; one is a faster runner; one is a better cook; one is an innate caregiver; one is more skilled at resolving conflicts than the other. Within a group larger than two, differences abound, inevitably. But the rules that hunter/gatherers developed are effective at minimizing the impacts of those differences on group dynamics. If one of the clan was highly skilled at finding and harvesting fruit but refused to share; or if one of the clan refused to offer firewood when the rest of the clan had none, those acts of selfishness could easily generate resentment and distrust with the group. If one of the clan assumed personal power over others and tried to make unilateral decisions that would affect the group or boss people around, that too could easily result in a divided group. If an argument between two individuals was not resolved, or was not resolved with the agreement and awareness of the whole clan, the group could split into divisive factions.

The clan increases its odds of survival when its individuals work together. Individuals increase their odds of survival when they work together as a group. The rules that hunter/gatherers evolved were applied for a reason: they are effective at minimizing inequality within the clan population and the disharmony inequality brings. Inequalities exist, but by not letting them enlarge into inflated egos, power imbalances and unequal allocation of resources, those inequalities were non-divisive.

As agriculturalists, we can learn from this. The lesson is to design and enforce public policy to minimize inequality between all individuals of the population. We have to create the cultural environment that nurtures our success. Human nature may tend toward wanting power, being selfish and finding fault with another person, but it is possible to minimize the expression of these tendencies through social pressure, and minimizing those expressions helps the clan. We are still the same organism we were 10,500 years ago. When we work together and support each other, we all do better. In the bigger picture the biological goal of long-term survival is still relevant.

A large population is more open to a disharmonic culture than a small one. In a large population, individuals cannot have a close relationship with all others and those who are not close are susceptible to discrimination. A large population naturally seeks a leader, an individual to make decisions and organize actions when needed, and with a leader comes hierarchy and power imbalances. Discrimination, hierarchy and power are all cultural aspects that hunter/gatherer policies worked to counter because of the disharmony they bring. In our large populations, it is much more difficult to design policies to generate harmony, but the cultural structure hunter/gatherers designed provides the template we can use on our population.

Problems of the environment are largely physical in nature. Water is available for sustenance or it is not. Industrial toxins are in our food or they are not. The ratio of gases in our atmosphere supports our lives on earth or it does not. Species are diverse enough to maintain a healthy ecosystem or they are not. However, problems of our culture are mostly a function of human nature, and our nature determines the policies we set and the decisions we make. We can decide to allocate resources equitably, or not. We can let power accrue to an individual or a sub-population, or not. We can determine that justice should be evenly distributed to the whole population, or not. This is the cost/

benefit analysis hunter/gatherers must have worked through and that influenced the structure of their cultural rules. Their clan populations were very small, 50 individuals and fewer, so policies were undoubtedly easier to apply, but the point is that now, with our large regional populations, it would be helpful to see that we have the ability to determine policies that would generate harmony with our group instead of disharmony and dissention. A divided population is an optional condition, not inevitable.

A Bigger Lesson

Besides the lessons we can apply to our relationships with the environment and our relationships with each other, we can look for a broader lesson. Historically, going back as many as 10,500 years, we developed and applied technology and grew our population in reaction to being afraid. As hunter/gatherers, it was probably a reasonable fear – if there was a neighboring population who would have willingly over-run your clan and you had nowhere to escape to, the concern for maintaining life was real. In this case, a group would use all tools at its disposal to survive. Other organisms, lacking an ability to make technology or to escape, would suffer a decrease in their population. Humans could use their evolved abilities to domesticate crops and grow their own food, build permanent structures, learn to create technology and grow their population as a way to defend against being overrun.

This step seems to have started a pattern. Our abilities to apply technology allow our population to grow, and the growing population then finds itself in the position of being hungry and still at risk of being overtaken by neighbors, so we feel the pressure to evolve our technology and grow more food. This perpetual cycle – a growing population needing more resources; more resources prompting population growth; population growth needing yet more resources – has continued to right now, this very moment.

But this pattern begs the question: is it helpful for us to respond to the fear of survival with growing our population and evolving our technology? Is our population better off with 7.5 billion people, widespread pollution, depleting resources and deep divisions between sub-populations of our species? We have replaced our wood and stone clubs with stunningly deadly arms that can obliterate millions of people in a flash – do these really help us? We have responded to fear with short-term fixes that are detrimental to us in the long-term. What really matters here? We make decisions and act based on immediate fear and panic, but those actions are diminishing our long-term prospects.

Our ancestors probably could not see the long-term problems that came with the trends they started, but we can. For our own long-term health, it behooves us to be aware of the impacts we are having on the environment and on each other, and act not out of fear but for our best long-term interests. Humans, like other organisms, tend toward ease and tend away from pain, but our unique abilities to create ease in the short-term are hurting us in the long-term. It is easier to drive a car than to walk or to ride a bike, but that ease of transportation takes a toll on our bodies and in the creation of long-term pollution. It is easy to turn up a thermostat and be warm all winter, but the environmental cost is enormous. It is easy and fun to fly to a distant country for a vacation, but the cost of that luxury is enormous for our long-term future. It is desirable to want safe neighborhoods, a good education, secure jobs and representation in government, but do we acquire those at the cost of depriving others in our population of those same things?

It really behooves us to find the pulse of our attitude and the differences between what we want and what we need to survive. Simply wanting something is not enough to justify taking it if the cost is long-term and deleterious to others.

Whether we recognize it or not, there is pressure from the natural world for *Homo sapiens* to maintain a population below the natural upper limit. Disease epidemics, shortages of resources, difficulty growing enough food to support a population, difficulty in resolving differences between sub-populations, cultural dissention, public violence and political disharmony are all symptoms of a population struggling to manage its relationship with the environment and it's relationships with each other. Unchecked, those problems would bring a shock to our population and a decrease in its size. A large population imparts stress on the system that would support a population existing below the natural upper limit. If our population remains above the natural upper limit, these symptoms will stay with us. The risk is of a population collapse, as the symptoms of dysfunction result in a population unable to provide what it needs to function for the long-term.

The Difficulty of the Goal

Hunter/gatherers give us the example of what a truly sustainable human lifestyle looks like. They lived it. Recognizing that we are agriculturalists and have been for over 10,000 years begs the question: could we really go back to living like our hunter/gatherer ancestors? Can we gather up and re-contain the contents of Pandora's box and not imagine, develop, apply and evolve technology, making our lives easier and therefore prompting a larger population? In a profound way, this would be denying our biological evolution: the physiological traits that we have evolved and the cognitive structure that defines how we see the world. This would be difficult to do, but if we want to resolve our environmental and cultural issues and dodge the inevitability of a population collapse, this is what we need to do.

How do we use our skills, our knowledge, the deep understanding and self-awareness we have as a species, to keep changing, to move forward in a direction that makes us healthy for the long-term? We do not need to look at the goal of living like our ancestors as being a step backwards – that is not a constructive perspective and not the point. Shifting our energies to make changes to our lifestyle that makes us viable for the long-term is moving in a forward direction. This is an extraordinarily difficult policy problem, but it is precisely what we need to face.

We react to many of the issues that negatively affect us – diseases, public violence, political dysfunction, shortages of resources, pollution of our water, air and soil – as specific problems to be solved. And this is appropriate, we should work to resolve immediate difficulty. But we also need to recognize that these problems emerge where and when they do because a set of variables created them. Widespread disease does not happen by chance. Concentrated power tends to corrupt. When a group is denied the resources of a culture, they will be frustrated. To solve these problems for the long-term, we truly need to address the variables that caused them.

Natural selection has bestowed on us a unique physiology that puts us in a spot of having tremendous power to change how we live. We made a significant change at the Neolithic Revolution. But in recognizing now that there are long-term negative consequences to that lifestyle, and that population collapse is virtually inevitable if we continue on this path, we need to figure out how to decrease the size of our population, stop using complex technology, and be satisfied with the lifestyle that comes with those changes. How we do this, what form our policies take, how we get to this point are difficult questions. We need to chart a course in unknown territory.

But, we can relearn what our ancestors knew, observe how they lived and absorb the attitude and the concepts they lived by. As a population composed of individuals, each individual has an impact on the whole. Each person who chooses to change to a lifestyle geared for long-term survival influences the whole population in that direction. The policies our ancestors applied in relating to their environment and to each other were designed for long-term viability. We are smart enough to do the same.

References

ACIA. 2005. *Arctic Climate Impact Assessment: ACIA Overview report* Cambridge University Press. 1042 pp. http://www.acia. uaf.edu

Agnoletti, Mauro and Simone Neri Serneri. 2014. *The Basic Environmental History* Cham, Switzerland: Springer.

Alacevich, Michele and Anna Soci. 2018. *Inequality: A Short History* Washington, D.C.: Brookings Institution Press.

Allsopp, Michelle and Richard Page, Paul Johnston, David Santillo. 2009. *State of the World's Oceans* Springer Publishing. springer. com

Alvaredo, Facundo and Anthony B. Atkinson, Thomas Piketty, Emmanuel Saez. 2013. "The Top 1 Percent in International and Historical Perspective" *Journal of Economic Perspectives* 27(3): 3-20.

Alvaredo, Facundo and Lucas Chancel, Thomas Piketty, Emmanuel Saez, Gabriel Zucman. 2017. *World Inequality Report* World Inequality Lab. https://wir2018.wid.world/

Ammerman, Albert J. 1975. "Late Pleistocene Population Dynamics: An Alternative View" *Human Ecology* 3(4): 219-233.

Armelagos, George J. and Alan H. Goodman, Kenneth H. Jacobs. 1991. "The Origins of Agriculture: Population Growth During a Period of Declining Health" *Population and Environment* 13(1): 9-22.

Arthur, W. Brian. 2015. *Complexity and the Economy* New York, New York: Oxford University Press.

Attwell, Laura and Kris Kovarovic, Jeremy R. Kendal. 2015. "Fire in the Plio-Pleistocene: the Functions of Hominin Fire Use, and the Mechanistic, Developmental and Evolutionary Consequences" *Journal of Anthropological Sciences* 93: 1-20.

Barnes, Ethne. 2005. *Diseases and Human Evolution*. Albuquerque, New Mexico, University of New Mexico Press.

Beck, Benjamin B. 1980. Animal Tool Behavior: The Use and Manufacture of Tools By Animals New York, New York: Garland Publishing.

Belfer-Cohen, Anna. 1991. "The Natufian in the Levant" *Annual Review of Anthropology* 20: 167-186.

Binford, Lewis R. 1983. *In Pursuit of the Past: Decoding the Archaeological Record* New York, New York: Thames and Hudson.

Borrell, Ferran and Aripekka Junno, Joan Antón Barceló. 2015. "Synchronous Environmental and Cultural Change in the Emergence of Agricultural Economies 10,000 Years Ago in the Levant" *PLoS ONE* 10(8): 1-19. DOI:10.1371/journal.pone.0134810

Boserup, Ester. 1981. Population and Technological Change: A Study of Long-Term Trends Chicago: University of Chicago Press.

Bowman, David M. J. S. and Jennifer Balch, Paulo Artaxo, William J. Bond, Mark A. Cochrane, Carla M. D'Antonio, Ruth DeFries, Fay H. Johnston, Jon E. Keeley, Meg A. Krawchuk, Christian A. Kull, Michelle Mack, Max A. Moritz, Stephen Pyne, Christopher

I. Roos, Andrew C. Scott, Navjot S. Sodhi, Thomas W. Swetnam. 2011. "The Human Dimension of Fire Regimes on Earth" *Journal of Biogeography* 38: 2223-2236.

Braidwood, Robert. 1964. *Prehistoric Men* Glenview, Illinois: Scott, Foresman and Company.

Brennan Center for Justice: Conference Report. 2014. "The Governing Crisis: Exploring Solutions" http://www.brennancenter.org/sites/default/files/publications/dysfunction%20V7%2005%2012.pdf Brennan Center for Justice at New York University School of Law. https://www.brennancenter.org/

Bush, Vannevar. 1945. Science: The Endless Frontier: A Report to the President On a Program For Postwar Scientific Research Reprinted July 1960. Washington, D.C.: National Science Foundation.

Caldeira, Ken and Michael E. Wickett. 2003. "Anthropogenic Carbon and Ocean pH" *Nature* 425(September 25): 365.

Calhoun, John B. 1962. "Population Density and Social Pathology" *Scientific American* 206(3) 139-148.

Carneiro, Robert L. 1970. "A Theory of the Origin of the State" *Science* New Series 169(3947): 733-738.

Carneiro, Robert L. 1986. "On the Relationship Between Size of Population and Complexity of Social Organization" *Journal of Anthropological Research* 42(3): 355-364.

Cazenave, Anny and William Llovel, 2010. "Contemporary Sea Level Rise" *Annual Review of Marine Science.* 2:145-173.

Childe, V. Gordon. 1951. *Man Makes Himself* London, England: New American Library.

Chiles, James R. 2001. *Inviting Disaster: Lessons From the Edge of Technology* New York, New York: HarperCollins Publishers.

Cleveland, Cutler J. and Christopher Morris. 2014. *Handbook of Energy: Volume II: Chronologies, Top Ten Lists, and World Clouds* Waltham, Massachusetts: Elsevier.

Coale, Ansley J. 1974. "The History of the Human Population" *Scientific American* 231(3): 40-51.

Cohen, Joel E. 1995. *How Many People Can the Earth Support?* New York, New York: W. W. Norton.

Cohen, Mark Nathan. 1977. *The Food Crisis in Prehistory: Overpopulation and the Origins of Agriculture* South Braintree, Massachusetts: The Alpine Press.

Cohen, Mark Nathan. 1989. *Health & the Rise of Civilization* New Haven, Connecticut: Yale University Press.

Cooley, Heather and Newsha Ajami, Mai-Lan Ha, Veena Srinivasan, Jason Morrison, Kristina Donnelly, Juliet Christian-Smith. 2014. "Water Governance in the Twenty-First Century" In *The World's Water, Volume 8, The Biennial Report on Freshwater Resources* Edited by Peter H. Gleick. Washington: Island Press.

Cosmides, Leda and John Tooby. Jerome H. Barkow. 1992. "Introduction: Evolutionary Psychology and Conceptual Integration" In *The Adapted Mind: Evolutionary Psychology and the Generation of Culture* Edited by Jerome H. Barkow, Leda Cosmides, John Tooby. New York, New York: Oxford University Press.

Crow, Ben and Suresh K. Lodha. 2011. *The Atlas of Global Inequalities* Berkeley, California: University of California Press.

Deaton, Angus. 2013. *The Great Escape: Health, Wealth, and the Origins of Inequality* Princeton, New Jersey: Princeton University Press.

De Vos, Jurriaan M. and Lucas N. Joppa, John L. Gittleman, Patrick R. Stephens, Stuart L. Pimm. 2014. "Estimating the Normal

Background Rates of Species Extinction" *Conservation Biology* 29(2): 452-462.

Diamond, Jared. 1999. *Guns, Germs, and Steel: The Fates of Human Societies* New York, New York: W. W. Norton.

Diamond, Jared and Peter Bellwood. 2003. Farmers and Their Languages: "The First Expansions" *Science* 300(April): 597-603.

Diamond, Jared. 2005. *Collapse: How Societies Choose to Fail or Succeed* New York, New York: Penguin Group.

Diaz, Robert J. and Rutger Rosenberg. 2008. "Spreading Dead Zones and Consequences for Marine Ecosystems" *Science* 321(August): 926-929.

Dunbar, Robin I.M. and Richard Sosis. 2017. "Optimizing Human Community Sizes" *Evolution and Human Behavior* 39: 106-111.

Dyurgerov, Mark B. and Mark F. Meier. 2005. "Glaciers and the Changing Earth System: A 2004 Snapshot" Occasional Paper 58. Institute of Arctic and Alpine Research, University of Colorado, Boulder, CO. http://instaar.colorado.edu/uploads/occasional-papers/OP58_dyurgerov_meier.pdf

Ember, Carol R. 2014. "Hunter-Gatherers" In *Explaining Human Culture*. C. R. Ember editor. http://hraf.yale.edu/ehc/summaries/hunter-gather

Evelyn, John. 1661. Fumifugium: Or the Inconveniencie of the Aer and Smoak of London Dissipated. Together With Some Remedies Humbly Proposed By J.E. Esq; To His Sacred Majestie, and To The Parliament Now Assembled https://archive.org/details/fumifugium00eveluoft/page/26

Feinman, Gary M. 1995. "The Emergence of Inequality: A Focus on Strategies and Processes" In *Foundations of Social Inequality*

Edited by T. Douglas Price, Gary M. Feinman. New York, New York: Plenum Press.

Ferrari, Alize J. and Fiona J. Charlson, Rosana E. Norman, Scott B. Patten, Greg Freedman, Christopher J. L. Murray, Theo Vos, Harvey A. Whiteford. 2010. "Burden of Depressive Disorders by Country, Sex, Age, and Year: Findings From the Global Burden of Disease Study 2010" *PLoS Medicine* 10(11) doi:10.1371/journal.pmed.1001547

Flynn, Rob. 2007. "Risk and the Public Acceptance of New Technologies" In *Risk and the Public Acceptance of New Technologies* Edited by Rob Flynn, Paul Bellaby. Houndsmills, Basingstoke, Hampshire, England: Palgrave Macmillan.

Foley, Jonathan A. and Navin Ramankutty, Kate A. Brauman, Emily S. Cassidy, James S. Gerber, Matt Johnston, Nathaniel D. Mueller, Christine O'Connell, Deepak K. Ray, Paul C. West, Christian Balzer, Elena M. Bennett, Stephen R. Carpenter, Jason Hill, Chad Monfreda, Stephen Polasky, Johan Rockström, John Sheehan, Stefan Siebert, David Tilman, David P. M. Zaks. 2011. "Solutions For a Cultivated Planet" *Nature* 478 (20 October): 337-342. doi:10.1038/nature10452.

Freedman, Ronald and Bernard Berelson. 1974. "The Human Population" *Scientific American* 231(3): 31-39.

Gebauer, Anne Birgitte and T. Douglas Price. 1992. "Foragers to Farmers: An Introduction" In *Transitions to Agriculture in Prehistory* Edited by Anne Birgitte, T. Douglas Price. Madison, Wisconsin: Prehistory Press.

Gowdy, John. 1999. "Hunter-Gatherers and the Mythology of the Market" In *The Cambridge Encyclopedia of Hunters and Gatherers* Edited by Richard B. Lee, Richard Daly. Cambridge, United Kingdom, Cambridge University Press.

Gowlett, J. A. J. 2016. "The Discovery of Fire By Humans: A Long and Convoluted Process" *Philosophical Transactions B* 371: 20150164.

Grattan, J. P. and R. B. Adams, H. Friedman, D. D. Gilbertson, K. I. Haylock, C. O. Hunt, M. Kent. 2016. "The First Polluted River? Repeated Copper Contamination of Fluvial Sediments Associated With Late Neolithic Human Activity in Southern Jordan" *Science of the Total Environment* 573(2016): 247-257.

Hansen, Gerrit and Makiko Sato, Reto Ruedy, Ken Lo, David W. Lea, Martin Medina-Elizade. 2006. "Global Temperature Change" *PNAS* 103(39): 14288-14293. www.pnas.org/cgi/doi/10.1073/pnas.0606291103

Hansen, Gerrit and Dáithí Stone. 2016. "Assessing the Observed Impact of Anthropogenic Climate Change" *Nature Climate Change* 6(May): 532-537. DOI:10.1038/NCLIMATE2896

Harmand, Sonia and Jason E. Lewis, Craig S. Feibel, Christopher J. Lepre, Sandrine Prat, Arnaud Lenoble, Xavier Boës, Rhonda L. Quinn, Michel Brenet, Adrian Arroyo, Nicholas Taylor, Sophie Clément, Guillaume Daver, Jean-Philip Brugal, Louise Leakey, Richard A. Mortlock, James D. Wright, Sammy Lokorodi, Christopher Kirwa, Dennis V. Kent, Hélène Roche. 2015. "3.3-Million-Year-Old Stone Tools From Lomekwi 3, West Turkana, Kenya" *Nature* 521(May): 310-315.

Harriss, R. C. and C. Hohenemser, R. W. Kates. 1978. "Our Hazardous Environment" *Environment* 20: 6-15, 38-41.

Hayden, Brian. 1995. "Pathways to Power: Principles for Creating Socioeconomic Inequalities" In *Foundations of Social Inequality* Edited by Douglas T. Price, Gary M. Feinman. New York, New York: Plenum Press.

Hellemans, Alexander and Bryan Bunch. 1988. The Timetables of Science: A Chronology of the Most Important People and Events in the History of Science New York, New York: Simon and Schuster.

Henry, Donald O. 1985. "Preagricultural Sedentism: The Natufian Example" In *Prehistoric Hunter-Gatherers: The Emergence of Cultural Complexity* Edited by T. Douglas Price, James Brown. Orlando, Florida: Academic Press.

Hidaka, Brandon H. 2012. "Depression As a Disease of Modernity: Explanations for Increasing Prevalence" *Journal of Affective Disorders* 140: 205-214.

Hitchcock, Robert K. and Megan Biesele. 2000. "Introduction" In: *Hunters and Gatherers in the Modern World: Conflict, Resistance and Self-Determination* Edited by Peter P. Schweitzer, Megan Biesele, Robert K. Hitchcock. New York, New York: Berghahn Books.

Hoegh-Guldberg, O. and P.J. Mumby, A.J. Hooten, R.S. Steneck, P. Greenfield, E. Gomez, C.D. Harvell, P.F. Sale, A.J. Edwards, K. Caldeira, N. Knowlton, C.M. Eakin, R. Iglesias-Prieto, N. Muthiga, R.H. Bradbury, A. Dubi, M.E. Hatziolos. 2007. "Coral Reefs Under Rapid Climate Change and Ocean Acidification" *Science* 318 (December): 1737-1742.

Hong, Sungmin and Jean-Pierre Candelone, Clair C. Patterson, Claude F. Boutron. 1996. "History of Ancient Copper Smelting Pollution During Roman and Medieval Times Recorded in Greenland Ice" *Science* 272: 246-249.

Hublin, Jean-Jacques and Abdelouahed Ben-ncer, Shara E. Bailey, Sarah E. Freidline, Simon Neubauer, Matthew M. Skinner, Inga Bergmann, Adeline Le Cabeel, Stefano Benazzi, Katerina Harvati, Philipp Gunz. 2017. "New Fossils From Jebel Irhoud, Morocco and the pan-African Origin of *Homo sapiens*" *Nature* 546(June): 289-292.

Hunt, Gavin R. and Russell D. Gray, Alex H. Taylor. 2013. "Why Is Tool Use Rare in Animals?" In *Tool Use In Animals: Cognition and Ecology* Edited by Crickette M. Sanz, Josep Call, Christophe Boesch. Cambridge, United Kingdom: Cambridge University Press.

Inglehart, Ronald. 2016. Inequality and Modernization; Why Equality is Likely to Make a Comeback Foreign Affairs 95(1): 2-10.

IPBES. 2019. Summary for Policymakers of the Global Assessment Report on Biodiversity and Ecosystem Services of the Intergovernmental Science-Policy Platform on Biodiversity an Ecosystem Services Edited by S. Diaz, J. Settele, E. S. Brondizio, H. T. Ngo, M. Guèze, J. Agard, A. Arneth, P. Balvanera, K. A. Brauman, S. H. M. Butchart, K. M. A. Chan, L. A. Garibaldi, K. Ichii, J. Liu, S. M. Subramanian, G. F. Midgley, P. Miloslavich, Z. Molnár, D. Obura, A. Pfaff, S. Polasky, A. Purvis, J. Razzaque, B. Reyers, R. Roy Chowdhury, Y. J. Shin, I. J. Visseren-Hamakers, K. J. Willis, C. N. Zayas. IPBES secretariat, Bonn, Germany. 56 pages.

IPCC, 2014: Climate Change 2014: Synthesis Report. Contribution of Working Groups I, II and III to the Fifth Assessment Report of the Intergovernmental Panel on Climate Change [Core Writing Team, R.K. Pachauri and L.A. Meyer (eds.). IPCC, Geneva, Switzerland, 151 pp.

Kates, R. W. and Christoph Hohenemser, Jeanne X. Kasperson. 1985. "Introduction: Coping With Technological Hazards" In *Perilous Progress: Managing the Hazards of Technology* Edited by R. W. Kates, Christoph Hohenemser, Jeanne X. Kasperson. Boulder, Colorado: Westview Press.

Keeton, William T. 1972. *Biological Science* New York, New York: W. W. Norton and Company.

Kelly, Raymond C. 2005. "The Evolution of Lethal Intergroup Violence" *PNAS* 102(43): 15294-15298.

Kent, Susan. 1989. "And Justice for All: The Development of Political Centralization Among Newly

Sedentary Foragers" *American Anthropologist* 91(3): 703-712.

Kjeldsen-Kragh, Søren. 2007. *The Role of Agriculture in Economic Development: The Lessons of History* Gylling, Denmark: Copenhagen Business School Press.

Kremer, Michael. 1993. "Population Growth and Technological Change: One Million B.C. to 1990" *The Quarterly Journal of Economics* 108(3): 681-716.

Lambin, Eric F. and B. L. Turner, Helmut J. Geist, Samuel B. Agbola, Arild Angelsen, John W. Bruce, Oliver T. Coomes, Rodolfo Dirzo, Günther Fischer, Carl Folke, P. S. George, Katherine Homewood, Jacques Imbernon, Rik Leemans, Xiubin Li, Emilio F. Moran, Michael Mortimore, P. S. Ramakrishnan, John F. Richards, Helle Skånes, Will Steffen, Glenn D. Stone, Uno Svedin, Tom A. Veldkamp, Coleen Vogel, Jianchu Xu. 2001. "The Causes of Land-use and Land-cover Change: Moving Beyond the Myths" *Global Environmental Change* 11(2001): 261-269.

Leakey, Richard and Roger Lewin. 1995. *The Sixth Extinction: Patterns of Life and the Future of Humankind* New York, New York: Doubleday.

LeBlanc, Steven A. 2003. *Constant Battles: Why We Fight* New York, New York: St. Martin's Press.

Lee, Richard B. and Richard Daly. 1999. "Introduction: Foragers and Others" In *The Cambridge Encyclopedia of Hunters and Gatherers* Edited by Richard B. Lee and Richard Daly. Cambridge, United Kingdom: Cambridge University Press.

Livi-Bacci, Massimo. 2017. *A Concise History of World Population* West Sussex, United Kingdom: Blackwell Publishing.

Longman, Jack and Daniel Veres, Walter Finsinger, Vasile Ersek. 2018. "Exceptionally High Levels of Lead Pollution in the Balkans From the Early Bronze Age to the Industrial Revolution" *PNAS* 115(25): 5661-5668.

Markham, Adam. 1994. *A Brief History of Pollution* New York, New York: St. Martin's Press.

Martin, Paul S. 1973. "The Discovery of America" *Science* 179(4077): 969-974.

McGinn, Anne Platt. 2000. "Why Poison Ourselves? A Cautionary Approach to Synthetic Chemicals" *Worldwatch Paper 153*. Edited by Chris Bright. Worldwatch Institute.

McNeill, J. R. and William H. McNeill. 2003. *The Human Web: A Bird's-Eye View of World History* New York, New York: W. W. Norton.

McPherron, Shannon P. and Zeresenay Alemseged, Curtis W. Marean, Jonathan G. Wynn, Denne Reed, Denis Geraads, Rene Bobe, Hamdallah A. Bearat. 2010. "Evidence For Stone-Tool-Assisted Consumption of Animal Tissues Before 3.39 Million Years Ago at Dikika, Ethiopia" *Nature* 466(August 12).

Middleton, Guy D. 2017. *Understanding Collapse: Ancient History Modern Myths* Cambridge, United Kingdom: Cambridge University Press.

Milanovic, Branko. 2011. "A Short History of Global Inequality: The Past Two Centuries" *Explorations in Economic History* 48: 494-506.

Minnesota Department of Natural Resources. https://www.dnr.state.mn.us/

Naroll, Raoul. 1956. "A Preliminary Index of Social Development" *American Anthropologist* 58: 687-715.

Nash, Linda. 1993. "Water Quality and Health" In *Water in Crisis: A Guide to the World's Fresh Water Resources* Edited by Peter H. Gleick. New York: Oxford University Press.

Neckerman, Kathryn M. and Florencia Torche. 2007. "Inequality: Causes and Consequences" *Annual Review of Sociology* 33: 335-357. doi: 10.1146/annurev.soc.33.040406.131755

Nriagu, Jerome O. 1996. "A History of Global Metal Pollution" *Science* New Series 272(5259): 223-224.

Piketty, Thomas. 2014. *Capital in the Twenty-First Century* Cambridge, Massachusetts: The Belknap Press of Harvard University Press.

Pinker, Steven. 2011. The Better Angels of Our Nature: Why Violence has Declined New York, New York: Penguin Group.

Postel, Sandra 1999. Pillar of Sand: Can the Irrigation Miracle Last? New York: W.W. Norton.

Postel, Sandra L. 2000. "Entering An Era of Water Scarcity: The Challenges Ahead" *Ecological Applications* 10(4): 941-948.

Price, T. Douglas and James A. Brown. 1985. "Aspects of Hunter-Gatherer Complexity" In *Prehistoric Hunter-Gatherers: The Emergence of Cultural Complexity* Edited by T. Douglas Price, James A. Brown. Orlando, Florida: Academic Press.

Price, T. Douglas. 1995. "Social Inequality at the Origins of Agriculture" In *Foundations of Social Inequality* Edited by T. Douglas Price, Gary M. Feinman. New York, New York: Plenum Press.

Price, T. Douglas and Anne Birgitte Gebauer. 1995. "New Perspectives on the Transition to Agriculture" In *Last Hunters, First Farmers: New Perspectives on the Prehistoric Transition to Agriculture* T. Douglas Price, Anne Birgitte Gebauer, editors. Santa Fe, New Mexico: School of American Research Press.

Railey, Jim A. and Richard Martin Reycraft. 2008. *Global Perspectives on the Collapse of Complex Systems* Albuquerque, New Mexico: Maxwell Museum of Anthropology.

Redman, Charles L. 1978. The Rise of Civilization: From Early Farmers to Urban Society in the Ancient Near East San Francisco, California: W. H. Freeman.

Redman, Charles L. 1999. *Human Impacts on Ancient Environments* Tucson, Arizona: The University of Arizona Press.

Roebroeks, Wil and Paola Villa. 2011. "On the Earliest Evidence for the Habitual Use of Fire in Europe" *PNAS* 108(13): 5209-5214.

Rosanvallon, Pierre. 2016. "How to Create a Society of Equals: Overcoming Today's Crisis of Inequality" *Foreign Affairs* 95(1): 16-22.

Ruddiman, William F. 2003. "The Anthropogenic Greenhouse Era Began Thousands of Years Ago" *Climate Change* 61: 261-293.

Ryan, Peter G. and Charles J. Moore, Jan A. van Franeker, Coleen L. Moloney. 2009. "Monitoring the Abundance of Plastic Debris in the Marine Environment" *Phil. Trans. R. Soc. B* 364: 1999-21-012.

Sahlins, Marshall. 1972. *Stone Age Economics* Chicago, Illinois: Aldine-Atherton.

Samuelson, Paul A. and William D. Nordhaus. 2001. *Microeconomics* Boston: McGraw-Hill Irwin.

Sandom, Christopher and Sören Faurby, Brody Sandel, Jens-Christian Svenning. 2014. "Global Late Quaternary Megafauna Extinction Linked to Humans, Not Climate Change" *Proceedings of the Royal Society B* 281: 20133254.

Schwartz, Glenn M. 2006. *After Collapse: the Regeneration of Complex Societies* Tucson, Arizona: The University of Arizona Press.

Scott, Andrew C. and William G. Chaloner, Claire M. Belcher, Christopher I. Roos. 2006. "The Interaction of Fire and Mankind: An Introduction" *Phil. Trans. R. Soc. B* 371(20150162). https://doi.org/10.1098/rstb.2015.0162

Shelp, Ronald Kent. 1981. *Beyond Industrialization: Ascendancy of the Global Service Economy* New York, New York: Praeger Publishers.

Simmel, Georg. 1902. "The Number of Members As Determining the Sociology Form of the Group" *The American Journal of Sociology* 8(1).

Smil, Vaclav. 2017. *Energy and Civilization: A History* Cambridge, Massachusetts: The MIT Press.

Steffen, Will and Angelina Sanderson, Peter D. Tyson, Jill Jäger, Pamela A. Matson, Berrien Moore III, Frank Oldfield, Katherine Richardson, H. John Schellnhuber, B.L. Turner II, Robert J. Wasson. 2005. *Global Change and the Earth System: A Planet Under Pressure* Berlin, Germany: Springer.

Tainter, Joseph A. 1988. *The Collapse of Complex Societies* Cambridge, United Kingdom: Cambridge University Press.

UNEP 2005: "Marine Litter: An Analytical Overview" United Nations Environment Programme. http://wedocs.unep.org/bitstream/handle/20.500.11822/8348/-Marine%20Litter%2c%20an%20analytical%20overview-20053634.pdf?sequence=3&isAllowed=y

van der Leeden, Frits and Fred L. Troise, David Keith Todd. 1990. *The Water Encyclopedia* Chelsea, Michigan: Lewis Publishers.

Waters, Colin N. and Jan Zalasiewicz, Colin Summerhayes, Anthony D. Barnosky, Clément Poirier, Agnieszka Galuszka, Alejandro Cearreta, Matt Edgeworth, Erle C. Ellis, Michael Ellis, Catherine Jeandel, Reinhold Leinfelder, J. R. McNeill, Daniel deB. Richter, Will Steffen, James Syvitski, Davor Vidas, Michael Wagreich, Mark Williams, An Zhisheng, Jacques Grinevald, Eric Odada, Naomi Oreskes, Alexander P. Wolfe. 2016. "The Anthropocene is Functionally and Stratigraphically Distinct From the Holocene" *Science* 351(6269): aad2622-1 – aad2622-10.

Weisdorf, Jacob L. 2005. "From Foraging to Farming: Explaining the Neolithic Revolution" *Journal of Economic Surveys* 19(4): 561-586.

Wikipedia. *List of Countries By GDP Sector Composition* (accessed 11/14/2019) https://en.wikipedia.org/wiki/List_of_countries_by_GDP_sector_composition

Wilkinson, Richard and Kate Pickett. 2010. *Spirit Level: Why Greater Equality Makes Societies Stronger* New York, New York: Bloomsbury Press.

Wilson, Edward O. 2015. "Of Ants and Men" The quotation in text is from a video, produced by Shining Red Productions. https://www.pbs.org/show/eo-wilson-ants-and-men/

Wirth, Louis. 1938. "Urbanism As a Way of Life" *The American Journal of Sociology* 44(1): 1-44.

World Health Organization, United Nations Environment Programme. 2012. *State of the Science of Endocrine Disrupting Chemicals 2012* Edited by Åke Bergman, Jerrold J. Heindel, Susan Jobling, Karen A. Kidd, R. Thomas Zoeller. Geneva, Switzerland: United Nations Environment Programme and World Health Organization. https://www.who.int/ceh/publications/endocrine/en/

World Meteorological Organization Greenhouse Gas Bulletin, No. 14, November 22, 2018. https://library.wmo.int/doc_num.php?-explnum_id=5455

Wrangham, Richard and Rachel Carmody. 2010. "Human Adaptation to the Control of Fire" *Evolutionary Anthropology* 19:187-199.

Wrigley, E. A. 2010. *Energy and the English Industrial Revolution* Cambridge, United Kingdom: Cambridge University Press.

Zalasiewicz, Jan and Mark Williams, Alan Smith, Tiffany L. Barry, Angela L. Coe, Paul R. Bown, Patrick Brenchley, David Cantrill, Andrew Gale, Philip Gibbard, F. John Gregory, Mark W. Houn-slow, Andrew C. Kerr, Paul Pearson, Robert Know, John Powell, Colin Waters, John Marshall, Michael Oates, Peter Rawson, Philip Stone. 2008. "Are We Now Living in the Anthropocene?" *GSA Today* 18(2): 4-8. doi: 10.1130/GSAT01802A.1A

Acknowledgements

The author would like to express deep appreciation to individuals who provided support at different points in the path that culminated in the publishing of this book. Professors Debra Levison and Dick Levins, literary agent Laurie Harper, literary editor Susan Thurston-Hamerski, friend Andi Scott Dumas, Ian Graham Leask, Gary Lindberg and the wonderful staff at Calumet Editions all had a hand in making this book come about. As well, the impromptu conversations with friends, friends of friends, neighbors and professors, too numerous to list, all helped nudge and form the eventual shape of this project. The author is grateful to each of these individuals. This support being acknowledged, the author takes full responsibility for any errors or omissions in the content of this book.

About the Author

Life-long concerns with environmental problems plus a Bachelor degree in biology led David to the question of how humans could cause the environmental problems that we do, while every other species is constrained from creating these same problems. A Master's degree in Public Policy provided a grounding in economics and policy dynamics and an understanding of how problems are solved at the political level.

www.ingramcontent.com/pod-product-compliance
Lightning Source LLC
Chambersburg PA
CBHW030936090426
42737CB00007B/448